The German Energy Transition

Thomas Unnerstall

The German Energy Transition

Design, Implementation, Cost and Lessons

 Springer

Thomas Unnerstall
Stockstadt, Germany

ISBN 978-3-662-54328-3   ISBN 978-3-662-54329-0   (eBook)
DOI 10.1007/978-3-662-54329-0

Library of Congress Control Number: 2017942779

Based on the German language book: *Faktencheck Energiewende, Konzept, Umsetzung, Kosten – Antworten auf die 10 wichtigsten Fragen* by Thomas Unnerstall, © Springer-Verlag Berlin Heidelberg 2016. All Rights Reserved.

Printed on acid-free paper

This Springer imprint is published by Springer Nature
The registered company is Springer-Verlag GmbH Germany
The registered company address is: Heidelberger Platz 3, 14197 Berlin, Germany

Preface

The German energy transition—widely known under the term *"Energiewende"*—is a remarkable phenomenon. First, there are certainly very few political projects in a Western democracy, if any, that span a period of 40 years. Second, back in 2011 the *Energiewende* set itself goals that 20 years ago would have been deemed simply impossible: to fuel a heavily industrialized country practically without relying on oil, natural gas, fossil power plants or nuclear power plants. Third, across almost the entire spectrum of political parties as well as within German society, there is a very unusual consensus on this basic policy—and at the same time there have been intense, controversial debates on virtually every measure taken over the course of its implementation during the past five years. Finally, the *Energiewende* has probably received more worldwide attention than most other decisions by a German government in recent decades.

The latest advances in international climate policy—notably the Paris climate agreement of December 2015 calling for new energy policies around the globe—have further contributed to the course of the *Energiewende* being closely watched by many interested parties outside of Germany: by international institutions, governments, think tanks, NGOs, journalists, researchers, students of politics and of energy policy and business leaders.

Given this state of affairs, it is somewhat surprising that there are only very few book-length studies on the subject, with the exceptions being mostly collections of separate contributions on different specific aspects of the *Energiewende*. Furthermore, these contributions are often written from a particular political point of view or a particular scientific point of interest. In other words, there does not seem to be a professional monograph offering a comprehensive, systematic and impartial account of the German energy transition.

This book intends to fill this lacuna. It presents a complete, clearly laid out description of the concept and the current status of the *Energiewende* as well as a structured account of its actual and foreseeable costs. It follows a strictly non-political, sober approach, basing the assessment throughout on figures and facts. Thus, the book does not intend to convince the reader of any particular point of view towards the *Energiewende*; rather, it intends to give structured information, to create a reliable overview and to deliver rational arguments. It is written for the observers from abroad enumerated above, i.e. for all those with a professional or a scholarly interest in the German energy transition.

My approach to the topic will make it possible to identify strengths and weaknesses of the concept of the *Energiewende* and to reveal successes as well as mistakes and aberrations in its implementation. Consequently—as we all learn best from mistakes—my study is able to point out the most important lessons to be learned, for countries undertaking a similar energy transition, from the German experience.

The two intentions of the book—to develop a comprehensive picture of the *Energiewende* and to highlight the most important lessons for the future—necessitate two major qualifications. First, I concentrate throughout on the essential aspects and on the most important figures, deliberately leaving out technical, political and legislative details. Second, and more importantly, I only treat the electricity part of the *Energiewende*, deliberately leaving out the heating and the transportation sector—even though the German energy transition comprises all three main sectors of energy consumption: electricity, heating and transportation. The justification for this (aside from the consideration that otherwise the book would have been too extensive) lies in the following: ever since its official launch in 2010/2011, the actual political work on the German energy transition (as well as, consequently, the national debate and the international assessments) has focused on electricity issues: electricity generation, power grids, electricity markets, etc. Only now—notably in the wake of the German "Climate Protection Program 2050" presented by the federal government in November 2016—is the political focus slowly shifting towards heating and transportation.

The composition and the style of the book are in line with its intentions:

- Instead of longer continuous texts, I have often used concise bullet points for the main thoughts, aspects or facts.
- Most figures have been generously rounded, to allow for quicker comparisons.
- Throughout the book, I often pause to summarize the preceding considerations in a standout print form, and at the end of Parts I, II and III, I have included a summary and a conclusion in separate chapters.
- References have been kept short, as all essential figures and facts are available online.

This book is not a scientific work, but it does claim to present all truly relevant facts, figures and arguments pertaining to the German energy transition in the electricity sector. However, despite my efforts to this end, I cannot hope to have succeeded in all respects. Therefore, I will be grateful for criticisms and suggestions from my readers.

In closing, I wish to express my deep gratitude to my friends who have encouraged me to write this book and to my colleagues in the energy industry who have greatly helped me to enhance my knowledge and to balance my views on the issues treated here. In these contexts, I am particularly indebted to Dr. Ulrich Dieckert, Prof. Dr. Ulrich Parlitz, Dr. Jan L. Hagens; and to Prof. Peter Birkner,

Ralf Klöpfer, Dr. Christoph Müller, Harald Noske, Jörg Stäglich, Norman Villnow, Guido Wendt and Dr. Bernd-Michael Zinow. It is also a pleasure to thank my editor Barbara Fess for her important support of the project and for the constructive, swift collaboration.

Finally, I would like to express my sincere hope that the German energy transition, despite its shortcomings and peculiarities, may indeed serve as an inspiration to many people around the globe who strive for climate protection through the transformation of the energy system in their own countries. If this book can be a small contribution to foster their productive spirit and to ease their paths, I will be greatly rewarded.

Stockstadt, Germany   Thomas Unnerstall
January 2017

Contents

List of Abbreviations

BNA	Federal Grid Agency for Electricity, Gas, Telecommunications, Post and Railway (*Bundesnetzagentur*)
CCS	Carbon capture and storage (technology for CO_2 filtering and storage)
CGP	Cogeneration plant
CHP	Combined heat and power (cogeneration)
DSM	Demand-side management
EEX	European Energy Exchange
EnWG	German Energy Industry Act (*Energiewirtschaftsgesetz*)
ETS	EU emissions trading system
EU	European Union
GDP	Gross domestic product
GREA	German Renewable Energy Act (*EEG*)
GW	Gigawatt
GWh	Gigawatt-hour
INDC	Intended Nationally Determined Contributions (to climate protection)
kW	Kilowatt
kWh	Kilowatt-hour
mio	Million (10^6)
MW	Megawatt
MWh	Megawatt-hour
PV	Photovoltaic
RE	Renewable energies
ROI	Return on investment
TSO	Transmission system operator
TSS	Trade and service sector
TWh	Terawatt-hour
UBA	German Environment Agency (*Umweltbundesamt*)

Introduction: What Is This Book About?

1.1 Five Questions

In Germany, few political concepts have penetrated so deeply into the public consciousness as the *Energiewende*. This German term describes the transition to a sustainable energy economy and literally means "energy turnaround", although it is usually translated as "energy transition". And it is unlikely that there exists another political project—or political vision—that enjoys such a wide-ranging, fundamental consensus in German politics, business and society as the *Energiewende*.

Internationally, too, the *Energiewende* certainly appears to have received more attention than most other projects by the German government. Many countries respect, even admire, Germany for this project and its—in global comparison—very ambitious goals. The main reason for the *Energiewende*'s remarkable popularity is undoubtedly the fact that it can essentially be regarded as the German response to *climate change*, which is perceived as one of the greatest challenges, if not *the* greatest challenge, that humanity faces today.

In marked contrast to this picture, ever since the official launch in 2010/2011 Germany has struggled through years of ongoing and intense public controversy around the *Energiewende*, including repeated and severe criticism of the project—with the result that many citizens (while maintaining a positive attitude to the *Energiewende* itself) are now skeptical with respect to its implementation.

Similarly, seeing the way that the *Energiewende* has developed over the last 5 years, the mood within the global debate about this project has likewise become more skeptical. Earlier in the decade, the majority of comments ran along the lines of: "The *Energiewende* is very demanding, especially from a technological perspective, but the Germans can do it". Today the tune has often changed to: "Germany's approach just isn't going to work—the *Energiewende* is far too expensive and too inefficient" (and sometimes international observers venture so far as to say "The Germans must be crazy if they go on like that").

© Springer-Verlag GmbH Germany 2017
T. Unnerstall, *The German Energy Transition*, DOI 10.1007/978-3-662-54329-0_1

Looking closer, key critical statements made about the *Energiewende* in the course of Germany's public debate turn out to be the same as those made at the international level:

- The *Energiewende* will cost an incredible amount of money; it is too expensive for the German economy.
- The *Energiewende* places too high a burden on private households (electricity prices are too high).
- The *Energiewende* threatens the international competitiveness of German industry (electricity prices are too high).
- German CO_2 emissions have barely decreased despite the *Energiewende*, and, in any case, the effect of the *Energiewende* on climate change is practically zero.

Are these statements true or false? Is this criticism of the *Energiewende* justified or not?

The first essential purpose of this book is to present a comprehensive, systematic and impartial account of the *Energiewende*—on the basis of which these questions can be answered objectively, i.e., drawing on the relevant facts, figures and reliable arguments.

Even though the *Energiewende* is *in its design* the transition to a sustainable energy economy *in all three energy sectors*—electricity, heating and transportation—in this book we confine ourselves throughout to the electricity sector. The main reason for this is the fact that *in the implementation* of the *Energiewende* so far (i.e. in the past 5 years), most of the political efforts, measures and discussions, most of the public debate in Germany as well as (consequently) most of the international attention have been devoted to the electricity sector. In a nutshell, in the first years of the project, the *Energiewende* in Germany was essentially not an *energy transition*, but (only) an *electricity transition*.

(It is certainly justified to criticize this state of affairs, i.e. to demand of the energy policy that similar emphasis be placed on the heating sector and the transportation sector. It would also be possible to point out why it is, on the other hand, certainly understandable that the German *Energiewende* policy has been one-sided so far in the above sense. And one can ponder the question when/to what degree this has to change in the coming years. But such considerations go beyond the scope of the book.)

Hence, in the following we will use the term "*Energiewende*" to mean the German energy transition in the electricity sector.

In addition to the above-mentioned change in the attitude of many international observers towards the German energy transition, the last 5 years have seen a second global development that is important with respect to the aims of this book: a growing awareness among many governments and in the international community of the urgency of the climate problem, which led to the Paris Climate Agreement in December 2015. Through this agreement, all major countries have committed themselves to pursuing the so-called "2 °C target", i.e. to cut their CO_2 emissions by the middle of the century to a level close to zero. And despite their differing attitudes toward nuclear energy, almost all countries agree that this target should mainly be achieved by way of renewable energies.

But this means nothing less than that a vast number of countries must undergo a fundamental transformation of their own electricity sector within the next 35 years. And since the electricity sectors in many of these countries have characteristics similar to Germany's before its *Energiewende*—at least 60% share of fossil fuels, electricity generation concentrated in relatively few large power plants, mostly close to electricity consumption centres—the challenges for these countries are actually quite similar to those posed for Germany, despite all the differences in detail.

This raises the following key question: If an *Energiewende* as discussed in this book (i.e. an energy policy with the primary goal of comprehensively replacing fossil fuels with renewable energies in the electricity sector) is essentially indispensable, or in other words, *if the German energy transition project is to a certain extent a prototype for similar projects in other countries*—what lessons can these countries learn from the German example?

Now it is clear that each such individual energy transition project depends heavily on the specific economic, political and technical conditions in the country, and so each country must find its own path. Nevertheless, since learning from mistakes is the best way to learn, we believe it will be helpful and important for national energy transformation processes in other countries to avoid a number of defects in the German energy transition project; in other words, not to commit again what now, with the benefit of 5 years of experience, can reasonably be considered essential political mistakes in the course of the *Energiewende*.

Therefore, **the second essential purpose** of this book is to identify these mistakes, again on the basis of the systematic account of the *Energiewende* we will present here.

The compostion of the book emerges from the two purposes outlined above. In the first part, we provide a clearly structured description of the basic concepts and the concrete design of the *Energiewende* as well as an analysis of its inevitable consequences for the German electricity system. The second part looks at the current status, mainly comparing the planned milestones of the *Energiewende* goals with the actual data of 2015 and 2016. This allows us to identify successes as well as shortcomings and failures.

The third part is devoted to the crucial issue of the costs of the *Energiewende*, likely to be the most controversial and critical topic—in the national debate as well as in the assessment of international observers. We present a reliable overview of the key figures (without going into too much detail), also covering the period up to 2030, and we discuss the essential effects on the main stakeholders of the national economy, private households and businesses. On this basis, we develop a rough but solid estimate as to the cost of an energy transition in any industrialised country, in relation to its GDP.

At the end of the book, in the fourth part, we will then sum up the issues outlined above; that is to say, we will comment on the following **five questions:**

1. Is the *Energiewende* too expensive for Germany's national economy?
2. Can private households and businesses cope with the financial burden resulting from the *Energiewende*?

(continued)

3. What are the actual effects of the *Energiewende* on CO_2 emissions?
4. What are lessons to be learned from the design of the *Energiewende*?
5. What are lessons to be learned from the implementation of the *Energiewende*?

1.2   Units

In this book, apart from € = Euro, $ = US dollar and t = metric ton, we use only two (energy-related) units: kilowatt and kilowatt-hour.

Kilowatt (kW)
It is a measure of the power output or capacity (maximum power output) of a power plant or the power consumption by a device or customer.
   Typical power plant ratings are:

Rooftop PV system	5 kW
Wind turbine	2000–3000 kW = 2–3 MW (megawatt)
Conventional large-scale power plant	1 million kW = 1 GW (gigawatt)

   Typical power consumption ratings in the electricity sector are:

Light bulb	0.06 kW
Hairdryer	1 kW
Large industrial facility	10,000–100,000 kW
Germany total (peak)	80 million kW

In this book we have generally used the unit GW (gigawatt) = 1 million kW.

Kilowatt-Hour (kWh)
It is a measure of the quantity of electricity a power plant produces (e.g. in 1 year) or that a device or customer consumes (e.g. in 1 year), respectively.
   Typical quantities of electricity produced by a power plant are:

Rooftop PV system	5000 kWh per year
Wind turbine	3–5 million kWh per year
Conventional large-scale power plant	5–7 billion kWh per year = 5–7 TWh per year

   Typical dimensions of electricity consumption are:

Refrigerator	70–100 kWh per year
Typical German household	3000 kWh per year
Large industrial facility	50–500 million kWh per year
Germany total	600 billion kWh per year

In this book we have generally used the unit TWh (terawatt-hour) = 1 billion kWh.

Table 1.1 Primary energy consumption in Germany (in TWh) in 2015

Primary energy sources	Quantity of energy (TWh)	%	Import quota
Oil	1250	34	98%
Natural gas	780	21	91%
Hard coal	470	13	89%
Lignite	440	12	0%
Nuclear energy	280	7	100%
Renewable energies	465	13	0%
Others	15	0	0%
Total	**3700**	**100**	**70%**

[1]

Table 1.2 Metrics for primary energy consumption (PEC), 2015

	World	Germany
PEC	170,000 TWh	3700 TWh (= 2.2%)
GDP/PEC	€ 0.4/kWh	€ 0.8/kWh
PEC per capita	23,000 kWh per capita	45,000 kWh per capita

GDP gross domestic product (at 2015 prices); [1, 2]

1.3   Basic Facts and Figures: Germany

1.3.1   Energy Consumption

In 2015, Germany consumed a total of 3700 TWh of primary energy (i.e. energy sources that are converted into either electricity, heat or transportation; Table 1.1).

A brief comparison with the worldwide primary energy consumption is shown in Table 1.2.

Regarding the three main sectors of energy consumption—electricity, heat and transportation—the final energy consumption of about 2465 TWh in 2015 is broken down as shown in Table 1.3.

1.3.2   Electricity Consumption

The approximately 600 TWh of electricity produced per year in Germany in recent years (2011–2015) to meet domestic demand for electricity (i.e. not including electricity exports)—the so-called *gross electricity consumption*—is broken down between the major consumer groups as shown in Table 1.4.

A brief comparison with worldwide gross electricity consumption is given in Table 1.5.

The development of (gross) electricity consumption in recent decades is shown in Table 1.6.

Table 1.3 Final energy consumption in Germany by sector (in TWh) in 2015

Sector	Quantity of energy (TWh)	%
Electricity	520	21
Heating	1230	50
Transportation	715	29

Heating = excluding heat from electricity; transportation = excluding electricity-based transportation; [1, 3]

Table 1.4 Gross electricity consumption by consumer group in Germany (in TWh) in 2015

Consumer group	Electricity (TWh)	%
Industry	230	44
TSS	80	15
Private households	130	25
Public sector and others	80	16
Final energy consumption—electricity	**520**	**100**
Power line losses/power plants	75	
Gross electricity consumption	**595**	

TSS Trade and service sector; [1]

Table 1.5 Metrics for gross electricity consumption (GEC), 2015

	World	Germany
GEC	24,000 TWh	595 TWh (= 2.5%)
GDP/GEC	€ 2.8/kWh	€ 5/kWh
GEC per capita	3300 kWh per capita	7400 kWh per capita

GDP gross domestic product (at 2015 prices); [1, 2]

Table 1.6 Development of gross electricity consumption in Germany (in TWh)

1990	2000	2010	2015
550	580	615	595

[4]

Table 1.7 German CO_2 emissions (in mio t)

CO_2 emissions	1990	2000	2010	2015
Energy related	990	840	780	750
Other	60	60	50	50
Total	**1050**	**900**	**830**	**800**

[5, 6]

1.3.3 CO_2 Emissions

Over the last 25 years, German CO_2 emissions have developed as shown in Table 1.7.

Add to this further greenhouse gas of currently about 100 million tons.

Table 1.8 Energy-related CO_2 emissions in Germany by energy consumption sector, 2015

Energy sector	CO_2 emissions (mio t)	%
Electricity (including exports)	310	41
Heating	280	38
Transportation	160	21
Total	**750**	**100**

[6, 7]

Table 1.9 Metrics for CO_2 emissions, 2015

	World	Germany
CO_2 emissions	36,000 mio t	800 mio t (= 2.2%)
CO_2 emissions/GDP	0.54 kg/€	0.27 kg/€
CO_2 emissions per capita	4.9 t per capita	9.8 t per capita

[2, 5]

Table 1.10 Development of CO_2 emissions from electricity generation excluding electricity exports (in mio t)

1990	2000	2010	2015
360	330	305	270

Electricity exports assumed to stem mainly from hard coal power plants; [7]

The next table (Table 1.8) shows a breakdown of energy-related CO_2 emissions to the three key sectors of energy consumption.

Table 1.9 gives a brief comparison with global CO_2 emissions.

The development of CO_2 emissions from electricity generation (excluding electricity exports) in recent decades is shown in Table 1.10.

1.3.4 Energy Imports

As illustrated above, Germany currently imports around 70% of its primary energy sources; the average cost of these imports in recent years is shown in Table 1.11.

The cost development in recent decades is given in Table 1.12.

In 2015, the sharp decline in global prices for primary energies led to these costs falling again to a level of about € 60 billion.

1.4 Basic Facts and Figures: OECD Countries

1.4.1 Energy Consumption

In recent years, primary energy consumption and energy efficiency for a number of OECD countries looked like as shown in Table 1.13.

Table 1.11 Average annual import costs for energy between 2010 and 2014 (€ billion)

Primary energy source	Costs
Oil	61
Natural gas	18
Hard coal	5
Nuclear energy	0.3

[1, 8]

Table 1.12 Cost development of German energy imports (in € billion per year, averaged)

	1990–1999	2000–2004	2005–2009	2010–2014
Energy imports	Approx. 20	Approx. 35	Approx. 60	Approx. 85

[1, 8, 9]

Table 1.13 Primary energy consumption (PEC) and energy efficiency in the OECD in 2014/2015

Country	Primary energy consumption (1000 TWh)	Energy efficiency (GDP/PEC) ($/kWh)
Germany	3.7	1.05
USA	26	0.7
UK	2.1	1.3
Australia	1.5	0.75
Canada	3.0	0.5
Mexico	2.2	1.0
Japan	5.1	0.95
OECD	61	0.85

GDP gross domestic product, in $ on the basis of 2015 Purchase Power Parities; [10, 11]

Table 1.14 Electricity mix in the OECD in 2014, shares in %

Country	Fossil fuels	Nuclear energy	Renewable energies
Germany	57	16	27
USA	67	19	14
UK	60	19	21
Australia	85	0	15
Canada	22	18	60
Mexico	79	3	18
Japan	85	0	15
OECD	59	18	23

[10, 11]

1.4.2 Electricity Generation

Turning to electricity, let us look at the current electricity mix of OECD countries in Table 1.14.

Table 1.15 Gross electricity consumption (GEC) and electricity efficiency in the OECD in 2014/2015

Country	Gross electricity consumption (TWh)	Electricity efficiency (GDP/GEC) ($/kWh)
Germany	600	6.0
USA	4300	4.2
UK	350	7.7
Australia	250	4.4
Canada	640	2.5
Mexico	300	7.3
Japan	1000	5.1
OECD	10,800	4.8

GDP gross domestic product, in $ on the basis of 2015 Purchase Power Parities; [10, 11]

Table 1.16 Key figures on CO_2 emissions in the OECD, 2014/2015

Country	CO_2 emissions (mio t)	CO_2 emissions/GDP (kg/$)	CO_2 emissions/capita (t)	Development of CO_2 emissions since 2000
Germany	800	0.2	10	−11%
USA	5600	0.31	17	−7%
UK	440	0.16	7	−22%
Australia	390	0.35	16	+12%
Canada	570	0.36	17	−24%
Mexico	470	0.21	4	+22%
Japan	1200	0.26	10	+5%
OECD	12800	0.25	10	−4%

GDP gross domestic product, in $ on the basis of 2015 Purchase Power Parities; [10, 11]

1.4.3 Electricity Consumption

If we look at the gross electricity consumption and at electricity efficiency within the OECD, the emerging picture is shown in Table 1.15.

1.4.4 CO_2 Emissions

Finally, we list important figures relating to CO_2 emissions in the OECD (Table 1.16).

References

1. AGEB (2016) Energy consumption in Germany in 2015. http://www.ag-energiebilanzen.de
2. Weltenergierat-Deutschland e.V (2016) Energie für Deutschland 2015. http://www.weltenergierat.de

3. AGEB (2016) Evaluation tables on the energy balance 1990 to 2015. http://www.ag-energiebilanzen.de

4. AGEB (2016) Stromerzeugung nach Energieträgern 1990–2016. http://www.ag-energiebilanzen.de

5. UBA (2016) Submission under the United Nations Framework Convention on Climate Change and the Kyoto protocol. http://www.umweltbundesamt.de

6. UBA (2016) Treibhausgasemissionen in Deutschland seit 1990 nach Gasen. http://www.umweltbundesamt.de

7. UBA (2016) Entwicklung der spezifischen CO_2-Emissionen des deutschen Strommix in den Jahren 1990–2015. http://www.umweltbundesamt.de

8. BAFA (1991) Aufkommen und Export von Erdgas sowie die Entwicklung der Grenzübergangspreise ab. http://www.bafa.de

9. BMWi (2013) Energie in Deutschland – Trends und Hintergründe zur Energieversorgung. http://www.bmwi.de

10. http://www.iea.org

11. http://www.oecd.org

Part I

The German Energy Transition: What Is Driving It?

Targets, Motives, Framework Conditions, Systemic Consequences

This first part of the book strives to comprehensively describe and explain the *Energiewende* in Germany.

The *Energiewende* is a political project launched, in its present form, during 2010/2011 by the German Federal government. The project is characterized by a very *long-term design*. It defines quantitative, verifiable targets in the energy sector for the year 2050.

In the spectrum of political issues and programmes, the *Energiewende* thus has a very unusual position. What other fields of politics have defined quantitative targets for 2050 or even 2030? It also far exceeds legislative periods and terms of office of individual federal governments and possibly the duration of some political party programmes. Taken seriously, it is therefore probably more appropriate to describe the *Energiewende* as a central *project of German society* in which the respective federal government plays the part of the responsible manager.

Given this background, what we do not want to do here is to explain the *Energiewende* from a *historical perspective:* what were the political origins, and what were the actual political considerations of the federal government in 2010/2011 that led to the design of the *Energiewende*? Instead we want to focus on the *systematic perspective*: what rational arguments can be presented in support of the design of the *Energiewende* that are sustainable in the long term?

This first part of the book also aims to provide a logical structure for the many buzzwords relating to the *Energiewende*—CO_2 emissions, renewable energies, storage, grid expansion, costs, blackout risk, rising electricity prices, falling prices on the energy exchange, falling profits of energy companies, nuclear energy phase-out and many more. Thus, our aim is not only to describe the *Energiewende* but also to *structure it conceptually*:

– What are the direct quantitative **targets** for the *Energiewende*?
– What political reasoning can be regarded as its deeper, long-term rational **motives**?

- What other principles of German politics and society play an essential role in the *Energiewende*, i.e. what is the **framework** within which it should be implemented?
- What **consequences** for the electricity system and for the electricity economy in Germany can be expected in the medium and long term, if the *Energiewende* is implemented consistently and systematically?

In summary, this first part is thus a systematically composed presentation, a *rational reconstruction* of the *Energiewende*. In keeping with the underlying concept of the book, we will limit ourselves to the essential aspects.

As is the case throughout the book, the figures given will typically be generously rounded so as to not burden the reader with unnecessary details and to illuminate the key structures as clearly as possible.

Three Targets of the Energy Transition: Description

The term *"Energiewende"* has a longer history in the political debate in Germany; it was first used in 1980 when it primarily meant the abandonment of nuclear energy and oil in the energy supply. Since then it has gone through several iterations. Generally speaking, the term *"Energiewende"* is mostly used today to refer to the transition in energy supply from fossil fuels and nuclear energy to renewable energies, in all three major energy-consuming sectors: electricity, heat and transportation.

In this book we use the term *"Energiewende"*—as already stated in the introduction—as an umbrella for (the key elements of) the federal government's energy policy since June 2011 *relating to the electricity sector*.

The *Energiewende* in this sense is characterized by **three pivotal targets**:

- Shutdown of nuclear power plants—by 2022
- Expansion of renewable energies in electricity generation—to at least 80% in 2050
- Increase in electricity efficiency—at a growth rate of around 1.6% per year

We will explain these three targets in more detail below. The *specific target figures* indicated are mainly taken from the "Lead Study 2011" [1] for the Federal Ministry for the Environment, scenario 2011A. The extent to which these targets *for 2015* have actually been achieved will be the main subject of the second part of this book.

© Springer-Verlag GmbH Germany 2017
T. Unnerstall, *The German Energy Transition*, DOI 10.1007/978-3-662-54329-0_2

2.1   Shutdown of Nuclear Power Plants

Germany's nuclear power plants, of which there were 18 at its peak, were built mainly during the 1970s and 1980s. Between 2000 and 2010 they generated a power output of about 20 GW and produced on average about 160 TWh of electricity each year [2]. Nuclear power plants thus provided approx. 20% of the required power output and approx. 25% of the required electricity in Germany.

Originally—from the 1950s to the mid-1970s—the peaceful use of nuclear energy was broadly accepted in Germany, even massively encouraged politically and supported by way of substantial subsidies. Since the late 1970s, however, the use of nuclear energy in power plants has been among the most violently and controversially debated topics not only within German energy policy but also within the overall political debate in Germany.

This is primarily a German phenomenon although other countries have had, and continue to have, similar discussions (e.g. Sweden, Switzerland and Italy). However, generally speaking these debates are not as significant within the political agenda.

The primary reason for the political debate surrounding the use of nuclear energy is the conflicting assessments of the risks posed by nuclear power plants for the present generation and by the radioactive waste they produce for future generations. Yet nuclear power plants indisputably have their advantages: in particular the avoidance of CO_2 emissions and the low cost of generating electricity in *existing* nuclear power plants. Therefore, the position one takes on nuclear power plants depends on how one weighs these benefits against the risks mentioned above. Ultimately this makes it a matter of value judgement.

The *Energiewende* target "shutdown of nuclear power plants" states, more specifically, that the remaining nuclear power plants are to be gradually shut down until 2022 (Table 2.1). Seven nuclear power plants with approximately 8 GW capacity were already closed in 2011, by order of the federal government, in the wake of the nuclear power plant accident in Fukushima, Japan.

This book does not address the various secondary issues involved: decommissioning of nuclear power plants, the search for locations to dispose of radioactive waste, allocation of the costs thus incurred, etc.

Table 2.1 Planned number of active nuclear power plants in Germany, 2000–2022

Year	2000	2010	2015 (planned)	2020 (planned)	2022 ff (planned)
Number	18	16	8	6	0

[1, 3]

2.2   Expansion of Renewable Energies in Electricity Generation

The term "renewable energies" (RE) is used to refer to energy sources that are present on the earth due to natural circumstances and are available, independently of any anthropogenic use, over long periods in the same manner. Since their use does not exhaust the natural resources of the earth, they are *sustainable*, and they are *CO2-free*. For electricity generation in Germany today, the available RE are mainly:

- Electricity from hydropower
- Electricity from wind power
- Electricity from sunlight (photovoltaic energy = PV)
- Electricity from (renewable) biomass

In the longer term, and depending on advances in technology, there could well be other forms to come, such as electricity generation through tides, waves, ground heat and others.

The RE used to generate electricity have been massively expanded in Germany since 2000, mainly through the funding mechanism of the German Renewable Energy Act (*Erneuerbare-Energien-Gesetz*, GREA) (Table 2.2).

As such, this part of the *Energiewende* has been going on for around 15 years already.

As a crucial part of the *Energiewende*, the explicit target has been set to expand the share of RE to at least 80% of the electricity consumed in 2050. The key milestones set in between are listed in Table 2.3.

Table 2.2 Electricity production from the major renewable energy sources, 2000, 2010 and 2016 (in TWh)

	2000	2010	2016
Wind	10	38	77
PV	0	12	38
Biomass	3	34	52
Water	25	21	21
Total	**38**	**105**	**188**

[2]

Table 2.3 Planned expansion of RE in Germany (in % of gross electricity consumption)

2000	2010	2015 (planned)	2020 (planned)	2030 (planned)	2050 (planned)
7	17	>26	>35	>50	>80

[1, 2]

2.3   Increase in Electricity Efficiency

The term "energy efficiency" generally describes the ratio of energy consumption to benefit achieved, specifically, the rate of energy consumption to achievement of a certain economic output. A commonly used measure of a country's energy efficiency is the gross domestic product (GDP) achieved per kilowatt-hour of gross energy consumption in €/kWh (also called energy productivity).

With respect to electricity, the focus of this book, we will use the term electricity efficiency. An increase in electricity efficiency thus means—to put it simply— maintaining (or even reducing) electricity consumption while increasing economic performance.

This is exactly what has happened in recent years: Electricity consumption in Germany has been stagnating in the last 15 years, while the economic output (GDP) has increased by 15% in real terms.

In its design of the *Energiewende*, the federal government has set a target for reducing gross electricity consumption by 25% by 2050, despite presumed further economic growth of about 1% per year. It is to be noted, however, that according to Scenario 2011A, in this decrease the electricity consumption for the production of hydrogen (2050: 110 TWh, primarily to serve as fuel in the transportation sector) remains unconsidered. This means that the increase rate for electricity efficiency must rise from about 0.5% per year between 2000 and 2010 to around 1.6% per year between 2010 and 2050 (Table 2.4).

2.4   Target State in 2050

What this design of the Energiewende means specifically in terms of installed RE plant capacity and RE electricity production will, of course, depend on the assumed electricity consumption in 2050. In the main scenario 2011A of the Lead Study 2011" [1] the three targets of the *Energiewende* and the respective milestones entail that electricity generation in Germany excluding electricity exports—i.e. gross electricity consumption—will develop up to 2050 as shown in Tables 2.5 and 2.6.

The balance in 2050 is also planned to include around 60 TWh of renewable electricity from abroad.

Table 2.4 Planned electricity efficiency (= gross domestic product/gross electricity consumption) (in €/kWh)

2000	2010	2015 (planned)	2030 (planned)	2050 (planned)
4.1	4.3	4.6	5.9	8.1

GDP at 2010 prices; 2010 = average of the years 2009–2011; [1, 4]

Table 2.5 Planned development of electricity generation in Germany in [1] (in TWh)[a]

	2000	2010	2015 (planned)	2030 (planned)	2050 (planned)
Nuclear energy	170	140	90	0	0
Fossil fuels	370	370	325	250	80
RE	40	105	170	300	430

[a] Excluding electricity generation for electricity exports; fossil energy = including others; 2030 adapted to current plans; electricity exports assumed to stem mainly from hard coal power plants; [1, 2]

Table 2.6 Planned development of electricity generation in Germany in [1] (in %)[a]

	2000	2010	2015 (planned)	2030 (planned)	2050 (planned)
Nuclear energy	30	23	15	0	0
Fossil fuels	63	60	56	45	16
RE	7	17	29	55	84

[a]Excluding electricity generation for electricity exports; fossil energy = including others; 2030 adapted to current plans; electricity exports assumed to stem mainly from hard coal power plants; [1, 2]

References

1. Leadstudy 2011 (Langfristszenarien und Strategien für den Ausbau der erneuerbaren Energien in Deutschland, 29.03.2012). http://www.dlr.de/dlr/Portaldata/1/Resources/bilder/portal/por tal_2012_1/leitstudie2011_bf.pdf
2. AGEB (2016) Stromerzeugung nach Energieträgern 1990–2016. http://www.ag-energiebilanzen.de
3. https://en.wikipedia.org/wiki/Nuclear_power_in_Germany
4. AGEB (2016) Energy Consumption in Germany in 2015. http://www.ag-energiebilanzen.de

Three Targets of the Energy Transition: Analysis

3

Looking at these three targets together, we can state the following:

- The three targets are not ends in themselves; they are pursued in order to achieve *superordinate goals* that we refer to as *motives* in this book to differentiate them on a conceptual level. We will describe these motives in the next chapter.
- The three targets of the *Energiewende* are mutually *independent*. Setting only two, or only just one, out of the three targets would certainly constitute a sensible energy policy.
- The crucial aspect of these targets is their *quantitative dimension*. This is what makes the *Energiewende* an ambitious and, so far, globally unique project. This is what largely defines the costs, policy tools and social conflicts associated with the implementation of the *Energiewende*.

In More Concrete Terms: An energy policy with the same targets *in terms of direction* but less ambitious quantitative design such as

- Shutdown of nuclear power plants by 2030
- Expansion of RE to 35% by 2030 and to 80% by 2050
- Increase in electricity efficiency by 1.0% per year

would satisfy the underlying motives just as well (only more slowly), but would definitely be much cheaper and easier.

> In other words, we have to distinguish between the basic concept—i.e. *target directions*—of the *Energiewende* and its concrete design, i.e. its specific quantitative targets and milestones. One can advocate for the former and consider the latter as being unnecessarily difficult and expensive.

© Springer-Verlag GmbH Germany 2017
T. Unnerstall, *The German Energy Transition*, DOI 10.1007/978-3-662-54329-0_3

Conversely,
the *Energiewende* would be conceivable with an even more ambitious design:

- Shutdown of nuclear power plants by 2018
- Expansion of RE to 80% by 2035
- Increase in electricity efficiency as before by 1.6% per year

Technically even these shorter-term targets are achievable—but this in turn would also have a significant impact on costs and policies of the *Energiewende*.

Four Motives of the Energy Transition: Description

4

For many years, energy policy played a relatively minor role in Germany. Time and again it would provide issues—the oil crisis (1973), the debate about the planned nuclear power plant Wyhl in Baden-Württemberg (1973–1983), subsidization of hard coal mining in Germany (1970–present), the debate about a new nuclear fuel rod reprocessing plant in Wackersdorf in Bavaria (1985–1989) and limitation of nitrogen oxide emissions particularly from hard coal-fired power plants—but these were not central to the political debate. (It is significant that the 1935 German Energy Industry Act was not substantially altered until 1998.)

The most enduring of these topics has certainly been the debate regarding nuclear energy, which was at least in part responsible for the emergence of the Green Party (today: Alliance '90/The Greens).

However, energy policy has become an established political and social issue in the last 15 years, when the following causal connection entered the international public consciousness and thus also that of German policymakers:

Use of fossil fuels $\rightarrow$ CO_2 emissions $\rightarrow$ climate change
$\rightarrow$ potential dramatic consequences of climate change

Despite all the differences in detail, there exists a broad consensus regarding this connection in German society and among most of the political parties in Germany.

A similar consensus has been achieved regarding nuclear energy following the events in Fukushima, Japan, in 2011. In the wake of this event, the CDU, CSU and FDP parties changed their energy policy programs. Since then, the belief that nuclear power is too dangerous to be used permanently in electricity generation has been a common basic motive in German energy politics.

The two central motives of German (energy) politics and of the *Energiewende* are therefore:

– Reduction of CO_2 emissions (limit climate change)

© Springer-Verlag GmbH Germany 2017
T. Unnerstall, *The German Energy Transition*, DOI 10.1007/978-3-662-54329-0_4

- Phase-out of nuclear energy (reduce the risks associated with the use of nuclear energy)

Two other, but comparatively less important, motives are (in order of importance):

- Reduction of dependence on fossil fuels (enhance energy security and improve trade balance)   .
- Promotion of innovation and thus export opportunities for German industry (enhance economic growth)

4.1   Motive 1: Reduction of CO_2 Emissions

Looking closer, this central motive in German energy policy and the *Energiewende* is actually ambiguous. Precisely what is driving the Energiewende policy:

1. The reduction of *national* CO_2 emissions
2. The reduction of *global* CO_2 emissions

It is clear that the problem of CO_2 emissions and ensuing climate change is by nature a *global problem*. Of course, reduction of *Germany's* national CO_2 emissions does contribute to the reduction of global emissions, but only to a negligible extent: Even a steady reduction, progressing so far as to completely eliminate German CO_2 emissions by 2050—i.e. a resounding success of German energy policy in this sense—would, all other things being equal, delay the climate change by a mere six months. In other words, it would be rational for German energy politics to focus on the reduction of *global* CO_2 emissions and to base the choice of political measures mainly on the criterion: how/where can we achieve— with a given amount of money and political effort—the greatest effect?

Despite of this, it is equally clear that German energy politics actually focuses almost exclusively on the reduction of *German* CO_2 emissions. The design of the *Energiewende*, the use of the GREA funds and the political measures are all dedicated to bring down CO_2 emissions from electricity generation in Germany.

In a nutshell, Germany's climate policy and energy politics—as actually the climate policy and energy politics of most developed countries—*should* be driven by the motive in the sense (2), but it is *in fact* driven by the motive in the sense (1).

What Does this Mean in Concrete Terms?
If the German government were to consistently act based on interpretation (2), this could potentially result in very different or additional political measures. Through the (at least partial) use of the funds flowing into the *Energiewende* for fostering the global spread of low CO_2 technologies, or for supporting energy transitions in other countries, it would probably be possible to affect global CO_2 emissions more significantly than through the *Energiewende* in its current design. However, regretfully, German policymakers have not yet addressed this issue

Table 4.1 Voluntary commitment to reducing greenhouse gas emissions (INDC) compared to the baseline year 2005, absolute reductions (in %); December 2015

Country	By 2025	By 2030	By 2050
USA	26–28	–	–
UK	–	38	–
Australia	–	26–28	–
Canada	–	30	–
Japan	–	25	–
Brazil	37	43	–
Germany	33	45	75

http://www4.unfccc.int/Submissions/INDC

Table 4.2 Voluntary commitment to reducing greenhouse gas emissions (INDC) compared to the baseline year 2005, reductions relative to GDP (in %); December 2015

Country	Until 2030	Until 2050
China	60–65	–
India	30–35	–
Germany	65–70[a]	90–93[a]

[a] Derived from the *Energiewende* targets and a nominal GDP growth of 2–3% per year; http://www4.unfccc.int/Submissions/INDC

seriously. Consequently, a more accurate wording for this first motive would be *reduction of CO_2 emissions in Germany*.

With respect to this motive, Germany has set itself indeed quite ambitious and far-reaching goals: compared to the baseline year 1990, reductions of 40% by 2020, of 55% by 2030 and of at least 80% by 2050.

The motive is, by and large, shared by the global community and most other countries, albeit with different conceptualizations and different priorities on the political agenda. Within Europe, it likewise plays a central role in EU energy policy and has, among other measures, led to the launch of the world's first CO_2 trading scheme (ETS), which we will revisit several times later on in this book.

The UN Climate Conference in Paris in December 2015 was an important milestone in the global context. At the conference, many countries expressly pledged to reduce their national CO_2 emissions (Tables 4.1 and 4.2); the urgency of the need to take action and the principal objectives of climate protection were set out in a binding agreement signed by almost all nations.

4.2   Motive 2: Phase-Out of Nuclear Energy

This second central motive for the *Energiewende* relates clearly directly to Germany and the German population.

However, the influence of Germany's energy policy is likewise somewhat limited. Even if this motive is very consistently implemented in the next years, the dangers of nuclear power are indeed going to be reduced, but will continue to exist to a large extent:

- Germany is surrounded by nuclear power plants in neighbouring countries, and nuclear accidents in these plants would also give rise to significant risks for Germany.
- The radioactive waste produced (and still being produced until 2022) through the operation of Germany's nuclear power plants must be disposed of in Germany, with corresponding potential and extremely long-term risks to the environment.

In contrast to the first motive, the second motive in this form is shared by only very few countries, and in virtually none of them it is awarded a similar priority. The few countries that are pursuing a policy of phasing out nuclear energy (e.g. Switzerland and Belgium) are planning to do so over significantly longer periods of time.

4.3   Motive 3: Reduction of Dependence on Fossil Fuels

It may well be surprising at first to see this motive listed as a separate topic. After all, one could object that the motive to "reduce CO_2 emissions" already implies a reduction in the use of fossil fuels, known to account for 85% of global as well as of German CO_2 emissions.

At least *historically*, however, the fact is that this subject predates the findings concerning climate change and is closely linked to the very first appearances of the term *"Energiewende"*.

More importantly for our purposes, this motive also plays a *systematic role* independent of the CO_2 issue. There are three reasons for this:

1. Since at present Germany has to import around 70% of its primary energy sources (see Table 1.1), "to reduce dependence on fossil fuels" automatically means "to reduce dependence on other—often politically unstable—countries" in relation to Germany's energy supply. In other words, to reduce dependence on fossil fuels means in fact to increase **energy security**. Both politics and society commonly perceive this as positive.
2. Furthermore, the aspect of **trade balance** is significant. The concern leading to this motive is that, as humanity continues to exhaust fossil fuels, the cost of importing these resources could continue to rise dramatically. This in turn would have negative consequences for Germany's trade balance and national economy. The development in recent decades justifies this concern (Table 1.12):
   - Between 1990 and 1999, Germany spent around € 20 billion per year on energy imports.
   - Between 2000 and 2009, the average spend was just under € 50 billion per year.
   - Between 2010 and 2014, the average spend was already approx. € 85 billion per year.

3. Finally, in this context it is important to mention the—certainly somewhat abstract—principle of a "**sustainable economy**". Though German consumption accounts for only around 2% of global consumption of fossil fuels, the fact is that every day our world consumes large quantities of irretrievable resources that took millions of years to form.

Conclusion
Even regardless of the issues of CO_2 and climate change, reducing dependence on fossil fuels is a traditional and significant motive of German politics. Accordingly, it is frequently brought up in discussions as an argument in favour of the *Energiewende*.

This motive plays a role in many other countries as well, although the manner in which it manifests varies widely. There are countries—even among the industrialized nations—where it is almost nonexistent. In contrast, in the USA this motive, specifically the aim to reduce dependence on *foreign* fossil fuels, is of central importance in (energy-related) politics. In particular, it has been a key factor in the spread of fracking. In other words, the USA has made and is still making considerable efforts to realize this motive (in acceptance of the significant potential environmental damage).

4.4 Motive 4: Promotion of Innovation/Export Opportunities for Germany's National Economy

This motive is again independent of the other three motives for the *Energiewende* and is repeatedly cited in discussions about the *Energiewende*.

Unlike the motives expounded so far, however, this motive is in no way energy specific: Similar effects can be achieved with very different instruments in other areas of politics. Moreover, even if we relate the matter to energy policy, no clear policy direction emerges. Exporting innovative or safety-optimized nuclear power plants or CCS technology (filter out CO_2 in fossil power plants) could also satisfy this fourth motive.

For these reasons, we will address this aspect only briefly in this book.

Four Motives of the Energy Transition: Analysis

5

5.1 Four Motives: General Overview

In summing up the relevant statements given in the preceding chapter, we can say that the basis for German energy policy thus defined for the next decades is practically unique in the world. No other comparable country is pursuing all these motives simultaneously.

This should come as no major surprise since Germany's consistent rejection of the continued use of nuclear energy is not replicated in many other nations. Additionally, *the two central motives*—to phase out nuclear energy and to reduce CO_2 emissions—*are diametrically opposed* and in this combination thus necessarily require a particularly challenging energy policy.

However, it is equally important to note that the *four motives reflect a clear, very broad consensus* among the German population, i.e. Germany's current energy policy is firmly anchored in society.

5.2 On the Relationship Between Motives and Targets

It is clear that the three targets of the *Energiewende* are sufficient to satisfy its four basic motives, at least in the longer term.

It is interesting, however, to pose this question the other way: If we accept the four guiding motives of German energy policy as a given (regardless of how they are rated and how important they are considered to be within the overall context of political convictions and motives), is the *Energiewende* then really a "TINA" (There Is No Alternative) case, at least in terms of its *direction*? Or would a substantially different energy policy be likewise able to satisfy the four motives?

On closer inspection, it quickly becomes evident that the *combination* of the motives is the deciding factor. For a combination of just *two* of the three main motives would give rise to alternative options:

© Springer-Verlag GmbH Germany 2017
T. Unnerstall, *The German Energy Transition*, DOI 10.1007/978-3-662-54329-0_5

- The two motives "reduction of CO_2 emissions" and "reduction of dependence on fossil fuels" could also lead to a policy of expanding nuclear energy.
- The two motives "reduction of CO_2 emissions" and "phase-out of nuclear energy" would also permit expansion of fossil electricity generation in conjunction with carbon capture and storage (CCS = filtering out CO_2 from the flue gases in power plants and subsequent storage of this CO_2 underground) without expanding renewable energies.
- Finally, the two motives "phase-out of nuclear energy" and "reduction of dependence on fossil fuels" (here understood to focus on foreign fossil fuels) alone could also bring about a strong, policy-based push towards expanding electricity generation from German lignite, possibly in combination with large offshore wind farms. Germany's lignite deposits are so extensive that it could cover 75% of the electricity generated in Germany for 50 years. However, realizing this would require a tripling of current levels of extraction and conversion of lignite into electricity, which would have a significant impact on land use and on CO_2 emissions.

Combining all three main motives, i.e. taking the guiding motives of current German energy policy as a basis, essentially leads us to the *Energiewende* concept in its present form, more specifically to the three targets (in terms of direction) of the German energy transition.

One alternative, conceivable in principle from a technological point of view, would be to implement a strategy of massive expansion of electricity generation from domestic lignite, combined with the widespread use of CCS technology—either exclusively or in combination with a limited expansion of RE. However, it must be noted that the CCS technology:

- Is not mature enough for such widespread use. (This could possibly be overcome through appropriate investment in research and development.)
- Cannot yet be assessed in financial terms
- Probably cannot meet the quantitative CO_2 targets
- Faces massive fundamental problems in terms of acceptance in the German society,

and so in the longer term, this theoretical option cannot really be considered realistic; in addition, it would not be sustainble.

Conclusion

Given the four motives of German energy policy, there is no reasonable alternative to the basic concept (i.e. the target directions) of the *Energiewende*.

5.3   Significance of the Quantitative Target Dimensions

The key message of the previous section is that, taking the four motives of German energy policy as a basis, the *Energiewende* is essentially a "TINA" case in terms of its direction.

However, we have already pointed out the need to distinguish between the three target *directions* of the *Energiewende* and its corresponding *specific quantitative targets*. Indeed, it is less the direction of the *Energiewende* than its quantitative, temporal dimensions that are responsible for most of the specific policy instruments and for a significant portion of the costs.

Therefore, we pose the question: Where do these explicit quantitative targets for the *Energiewende* come from? To what extent are they also a logical consequence of the four political motives and to what extent are they actually arbitrary and could be changed without discrediting the *Energiewende* as such?

Upon impartial analysis, this question must be answered twofold.

On the One Hand
In the global climate debate, there is an important basic thesis that is shared by not all, but by most climate scientists: the so-called 2 degrees Celsius goal (hereafter "2 °C goal"). The thesis is as follows: If we were able to limit climate change so that the global average temperature rises by less than 2 °C above pre-industrial levels, then we would have some sway over the consequences of climate change. Otherwise, i.e. if the average temperature increases by more than 2 °C, the consequences will be far more dramatic and barely manageable.

The "2 °C goal" served as the basis for global action in the Paris Climate Agreement of December 2015. The scientific experts are convinced that, if at all, this can only be achieved if the industrialized nations cut their CO_2 emissions by at least 80% by 2050 compared to 1990.

The German government has advocated for this goal, has supported it at international conferences and so, consequently, to reduce German CO_2 emissions by at least 80% by 2050 should be the guideline for its own political efforts. And it is precisely this point—the reduction in German CO_2 emissions by at least 80% by 2050 (with a simultaneous rapid phase-out of nuclear energy)—that can actually be considered the guiding principle in the *Energiewende* policy.

This principle thus formed the central condition to be met in the expert opinions commissioned by the German government in 2011 to design the path for the *Energiewende*. In fact, the path designed by the German government, in particular:

- The rate of increase in electricity efficiency
- The milestones in the expansion of renewable energies (2020, >35%; 2030, >50%; 2040, >65%; 2050, >80% share in electricity generation)

is derived from the scenarios outlined in these expert opinions.

In this sense the explicit quantitative targets of the *Energiewende* relating to electricity efficiency and the expansion of renewable energies can be considered to be conceptually sound.

Table 5.1 Conceivable alternative path for the *Energiewende*

Target dimension	Alternative path	*Energiewende* milestones
Shut down nuclear power plants	By 2030	By 2022
RE-share by 2020	>25%	>35%
RE-share by 2030	>35%	>50%
RE-share by 2040	>55%	>65%
RE-share by 2050	>80%	>80%

2 °C goal:

→ CO_2 emissions of industrialized nations = −80% by 2050
→ German CO_2 emissions = −80% by 2050
→ Scientific expert opinion with this condition
→ Possible paths to increasing electricity efficiency and expanding renewable energies
→ Quantitative targets/milestones for the *Energiewende*

On the Other Hand
Doubts are justified if we consider the last step of the above train of thought. There is more than one viable path to the desired target state in 2050, and good arguments for the German government's choice are difficult to find.

In more concrete terms: If the desired target state in 2050 consisting of:

– No nuclear power
– At least 80% of electricity generated from RE
– 25% less electricity consumption (excluding electricity for hydrogen production)

is deemed necessary, then a path towards this state as shown in Table 5.1 would be just as reasonable an option.

In the years from 2011 to 2016, and in the following years such an alternative dimensioning of the *Energiewende* targets would have led, or would lead, to policies that differ significantly from the measures actually taken or that are foreseeable.

Conclusion
The design of the *Energiewende*—i.e. its quantitative target dimensions that stand as defining characteristics—is on the whole well understandable. However, it also includes arbitrary elements as regards the speed of the transition, with quite serious consequences for Germany's specific energy policy, at the latest since 2011.

At first glance it may seem that with the three targets for the *Energiewende*—including their quantitative dimensions—and the underlying four basic political motives, the energy policy is largely fixed and that Germany's energy landscape in the next decades is quite clearly defined.

Looking closer, however, one observation in particular points to the fact that this is not the case. After long years of political infighting, since 2011 there has been broad agreement among the major political parties in Germany and a broad and stable consensus throughout society regarding the targets and motives of the *Energiewende*. Yet there are still intense, and not infrequently bitter, political and social discussions and controversies about the proper implementation of the *Energiewende*. How can this be?

Regardless, and from a purely objective point of view, within the constraints imposed by the three targets and four motives and on the basis of currently available technologies, there are in fact numerous potential ways of implementing the *Energiewende*. In other words it is possible to imagine quite different "*Energiewende* landscapes" for Germany.

Let us briefly outline two of these for clarification.

A Consistently *Central Energiewende*

In this variant of the *Energiewende*, from the very beginning (but at least from 2011 on), the state—i.e. specifically an authority equipped with appropriate powers—would have centrally planned the locations for the RE plants to be built, bundled multiple plants in locations selected for their optimal natural conditions, tendered the plants' construction and operation and thus created a system of large-scale renewable energy power plants. Simultaneously, and with the help of a national state-owned transmission system operator, the state would have built an advanced transport grid to distribute the renewable electricity thus generated to the consumption centres. In addition, in the next decades, this authority would also have built the necessary storage facilities. Furthermore, it would have aligned the structure of the fossil power plant fleet so that some of the current challenges on the financial side

© Springer-Verlag GmbH Germany 2017
T. Unnerstall, *The German Energy Transition*, DOI 10.1007/978-3-662-54329-0_6

and as regards CO_2 emissions (which are discussed in more detail in the second part of the book) would have been avoided from the outset, for example, by gradually replacing hard coal-fired power plants with gas-fired power plants.

A Consistently *Decentralized Energiewende*
In this variant of the *Energiewende*, a dream would come true that is still being dreamt in many towns and cities in Germany: the dream of energy self-sufficiency or at least—in line with the focus of this book—electricity self-sufficiency.

In fact, even with the technologies available today, it would be possible to make most of the smaller towns and cities (largely) self-sufficient in terms of electricity. All it would take is a suitable combination of local, small-scale wind farms, local biomass plants, PV systems, local cogeneration plants and local storage facilities. As for the major cities, this is more challenging but essentially just as feasible. Major cities would still need to be surrounded by large conventional power stations for a significant period of time, to be gradually supplemented and then replaced by countless photovoltaic, wind and biomass plants including storage at the periphery.

In the electricity system created by this variant of the *Energiewende*, the need for the transmission grid would gradually disappear, and there would be no need to construct new large power lines.

It Is Quite Clear Both these scenarios, i.e. both "energy landscapes", can satisfy the targets and motives of the *Energiewende*. Yet they are fundamentally different, and the "real" *Energiewende*—meaning the path which is actually pursued politically at this time, the actual energy landscape in Germany in 2016—again looks quite different. More accurately, it comprises some elements from scenario 1 and some from scenario 2.

Why is Germany not taking the first or the second path, even though both have significant respective advantages and would prevent quite a number of the current problems and debates?

The reason is, in a nutshell, scenario 1 ultimately means a planned economy and is therefore not compatible with the principle that in Germany the energy sector, and particularly electricity generation, is and should be governed by market economy rules; scenario 2 is extremely expensive and thus incompatible with the principle that energy should be affordable.

These principles, which basically guide the German government's energy policy (in addition to historical incidents and tactical considerations, which always play a significant role in how basic policies are specifically implemented), lead to Germany's choice of its path out of the numerous possible routes of an *Energiewende*.

Conclusion
The implementation of the *Energiewende* by the German government in its present form can be understood only by taking into account that there are other important, fundamental political principles relating to German energy policy that have so far not been addressed here. In this book, we refer to them as the *framework conditions of the Energiewende*.

The three essential framework conditions are:

- Security of supply
- Affordability/cost efficiency
- Market economy in the energy sector

Let us describe them consecutively.

6.1   Security of Supply

Security of supply in the electricity sector is defined as the continuous availability of electricity at all times. It is usually measured as the duration of average power outages per year, i.e. the statistical average time for which each electricity consumer is without electricity. In Germany, for many years this time has been about 15–20 min per year—a top value in Europe and worldwide. When it comes to the *Energiewende*, the framework condition "security of supply" means that the transformation of the energy landscape, and of electricity generation in particular, should proceed in such a way that the above value does not deteriorate—or at least not significantly.

There are other, secondary aspects of security of supply, especially so-called micro-blackouts (voltage fluctuations in the millisecond range), that we will not address in this book.

6.2   Affordability/Cost-Efficiency

The framework condition of "affordability" states that German energy policy has to be such that electricity remains "affordable" for private households as well as for companies. Of course, it is not really clear what "affordable" means in concrete terms. This framework condition therefore translates best into actual politics as the principle of *cost-efficiency*: of several possible future paths for the *Energiewende*, the path—or more cautiously one of the paths—should be selected that results in the lowest macroeconomic cost and therefore imposes the least financial burden on the national economy, that is to say on private households and companies.

It is probably obvious that realizing this principle in practice can be both challenging and controversial. The complexity of macroeconomic relationships and aspects often makes it difficult to calculate the cost of certain paths or specific measures in advance and in detail.

This framework condition, i.e. the question of affordability and cost efficiency within the *Energiewende*, plays an essential role in this book.

6.3   Market Economy in the Energy Sector

For the purposes of this book, the term "market economy" denotes the framework condition to German energy policy that it must comply with the basic framework for the energy economy that was established in and has been valid since 1998. This framework—mostly defined by the German Energy Economy Act (*Energiewirtschaftsgesetz*, EnWG) and a whole host of ordinances—states that the energy sector is subject to free competition. As a matter of principle, any stakeholder can build a power plant and offer the electricity to other stakeholders in the market. Any stakeholder can buy electricity in this market and offer it to (end) customers, such as industrial firms or private households. And every customer is free to choose from whom to buy electricity to meet their needs.

The only exception consists in the power grids, which form a natural monopoly and are therefore subject to state control in terms of operation and pricing.

In electricity generation, trading and sales (as well as other energy-related services such as energy consulting, contracting, etc.), a market economy has been in play only since 1998. For the 80–90 years prior, Germany's electricity economy was a monopoly economy in which all prices—and thus the construction of power plants in particular—were subject to state control.

The main impulse for the change to today's system came from Europe in the 1990s, i.e. from the EU. Cross-border competition within the EU was to be established, and so liberalization of the electricity markets in the member states became a mandatory prerequisite. Already this embedding of the German energy economy legislation into the European body of law, but also the importance of market economy as a fundamental pillar of the political and social order in Germany, results in the framework condition for the fundamental transformation induced by the *Energiewende* in the energy economy that it must leave the above principles (largely) intact.

Since the *Energiewende* is above all a *change in the electricity generation landscape* in Germany, in more specific terms the path for the *Energiewende* should be set out, as far as possible, such that every stakeholder is still permitted to build (or shut down) power plants, that the price of electricity from power plants is still set in the market and that the market, rather than a public authority, decides on the success, mode of operation and utilization of power plants.

Framework Conditions of the Energy Transition: Analysis

7

7.1 Tensions

Taking first a cursory look at the three framework conditions—security of supply, affordability/cost efficiency and market economy in the electricity sector—we note:

- Together with "environmental compatibility", the first two framework conditions—security of supply and affordability—form the famous energy policy triad in Germany, which is cited at the beginning of almost every brochure, on every website and in every speech given by the German government. As such they really do serve as fundamental guidelines, indeed as framework conditions, which all German energy policy needs to comply with. Or with which it must at least claim compliance. This triad emerged quite independently of, and much earlier than, the *Energiewende*. It has existed for decades, and every German government, whatever its constellation, has accepted it as a baseline and advocated for it.
- By contrast, the principle that energy policy, specifically the *Energiewende* policy, should operate within the conceptual boundaries of a market economy, in particular with respect to electricity generation, is more implicit and (as previously mentioned) originated more recently. Nevertheless, we can certainly assume that most major political players in Germany approve of this principle.

It is quite obvious, however, that the *Energiewende* is indeed in a tense relationship with all three framework conditions:

- It represents a challenge to the *security of supply* due to the dependence of RE electricity production on weather conditions that are in no way correlated with the demand for electricity to be met.

© Springer-Verlag GmbH Germany 2017
T. Unnerstall, *The German Energy Transition*, DOI 10.1007/978-3-662-54329-0_7

- It constitutes a challenge for *affordability* as well, since without it electricity would be noticeably cheaper today for most end consumers—mainly because the GREA surcharge (see Sect. 17.5) of € 0.0688 per kWh (2017) would not exist.
- Finally, it involves a significant government intervention in the electricity generation *market* per se.

These are tensions, not contradictions. They merely highlight the fact that an *Energiewende* policy that subordinates itself to the three framework conditions—as it is generally demanded, at least in an abstract manner, in the German political landscape and in German society—is an ambitious undertaking. It must seek the specific implementation path that (beyond its innate, unavoidable degree of impairment) least affects the principles of security of supply, affordability and market economy.

> In other words, in implementing the *Energiewende*, the German energy policy is required to keep tensions with the framework conditions to the unavoidable minimum inherent in the *Energiewende* project.

7.2   Is the Future of Germany's Electricity System Predetermined?

7.2.1   The Question

The thread of thought now leads us inevitably to the following question: If the *Energiewende* is wanted and the three framework conditions are accepted, to what extent does this *determine* the implementation path?

To Wit
If it is true that the targets and motives of the *Energiewende* alone still allow for several, very different approaches to implementation—thus permitting different future "energy landscapes"—to what extent do the three framework conditions then lead to *one* of these implementations, *one* path, *one* future energy landscape in Germany?

This question may seem academic. However, we can also pose the same question a different way: Are the continued intense political debates, even confrontations pertaining to the *Energiewende*—ongoing despite a broad consensus in politics and society with respect to its targets, motives and framework conditions—actually really necessary? That is to say:

- Do they originate from the fact that, even given these conceptual pillars of the *Energiewende*, there are still different options when it comes to concretely implementing it, from which a selection has to be made by taking into account yet other aspects and political preferences?

If so, we would conclude that the political disputes are primarily justified and inevitable.

– Or do they stem from the fact that, even though certain protagonists within the political debate agree with the consensus in the abstract, they nevertheless reject some of the inevitable consequences (for whatever reason); or to put it more clearly, do they stem from the *inconsistency* of certain protagonists?

In this case we would conclude that the political debates are in fact not justified, i.e. can be ascribed to the typical political power plays.

Hence our initial question is important for a better understanding of the political debates in Germany concerning various aspects of the *Energiewende*.

7.2.2   The Answer

Our answer to the question posed is as follows:

Taking the targets, motives and the three framework conditions as a basis, significant aspects of the future path of the *Energiewende* (and thus of the future of the German electricity system) are in fact largely predetermined.

Laying out these aspects is the task in the next chapter on "Systemic consequences".

7.2.3   Justified and Unjustified Discussions

However, there are also significant aspects where opinions may differ based on sound arguments in each case. These aspects include, for example, the definition of the market rules for conventional power plants over the next decade.

A more in-depth analysis of these discussions is beyond the scope of this book, but this much can be said here: The fact that there are often quite different views—even among scientists—on certain options for the further implementation of the *Energiewende* can essentially be explained by two factors:

– First, the consequences of a particular decision or approach (in particular the macroeconomic costs) can be difficult to determine clearly since they relate to complex markets, multifaceted relationships involving external factors and a variety of stakeholders.

– Second, even if all the consequences of a decision are relatively undisputed, there may be differences in emphasis, for example, in the weighting of the three framework conditions or of the motives.

> **Conclusion**
> In the course of the implementation of the *Energiewende*, there are (and will be) a number of quite controversial—and from the *perspective of individual stakeholders,* weighty—policy issues for which there is not the one and only right and consistent decision.

Therefore, it is both reasonable and understandable that differing points of view will persist in implementing the *Energiewende*, be it with respect to the objective impact assessment of certain options or be it with respect to priorities among its principles. The responsible policymakers will have to weigh these aspects to decide between the solutions in question. This is often tedious, gives rise to heated debates, creates an impression of chaos and discord and may even bear the risk of the *Energiewende* being seriously disavowed.

In fact, however, this is inevitable. The energy economy is too complex and far too intermeshed internationally and is also exposed to too many influencing factors, to justify the expectation of unanimity in all matters, even with a strong and honest underlying consensus on the fundamentals of the *Energiewende*.

(An unpleasant aspect here is that, in the public debate in Germany, those who advocate various solutions frequently accuse each other of "betraying the *Energiewende*" or even predict that the *Energiewende* would fail if the solution proposed by their opponent were implemented. That is pure polemics and so represents exactly what the other side is accused of: betraying the *Energiewende*.)

To Sum Up It is quite important to distinguish the issues and political decisions just considered from those for which the targets, motives and framework conditions do set out a clear direction for the *Energiewende*.

Accordingly let us illustrate which developments stemming from the *Energiewende*, which outlines of the future energy landscape in Germany *cannot* reasonably be brought into question, i.e. should not actually be the subject of political debate.

The next chapter is devoted to this task.

Systemic Consequences

8

The aim of this chapter is to bring out the inevitable consequences for the concrete energy landscape in Germany that will emerge due to the *Energiewende* as we have described it in terms of targets, motives and framework conditions. We refer to these as the *systemic consequences* of the *Energiewende*.

8.1 Types of Renewable Energy 1

If one of the three targets of the *Energiewende* is defined as "expansion of renewable energies to at least 80% of Germany's electricity generation", this does not yet constitute a statement about which of the currently available technologies in the field of renewable energy should be used here. So we will first address the question: Which renewable technologies should be expediently used in the *Energiewende*?

As of today, in Germany the following technologies can be considered viable in terms of technological maturity, permitting the performance of an RE plant to be largely predictable (in terms of power output, electricity production, service life, effects on the immediate environment, costs, etc.):

– Hydropower plants (running water)
– Biomass plants (biogas, wood, domestic waste)
– Wind turbines on land (onshore)
– Wind turbines at sea (offshore)
– PV systems

It is quite possible that there will be technological advances made during the next few decades that will allow the use of other types of renewable energy and so will play a role in the future implementation of the *Energiewende* (e.g. geothermal power plants, tidal power stations and others). However, it is obvious that the question posed at the beginning of this section must be answered regardless.

© Springer-Verlag GmbH Germany 2017
T. Unnerstall, *The German Energy Transition*, DOI 10.1007/978-3-662-54329-0_8

Concerning the renewable energy technologies of hydropower and biomass:

Hydropower Plants

Hydropower plants have been in use for decades (regardless of energy policy), i.e. as the only RE, they have always been an integral part of the electricity mix in Germany. However, years ago hydropower already—and here all the experts agree—essentially reached its natural limits of around 5 GW of output and 20–25 TWh per year (depending on weather conditions) of electricity production. It therefore has no role to play in the planned expansion of renewable energies and in the *Energiewende*.

Biomass Power Plants

Biomass power plants have been expanded from 1 GW in the year 2000 to almost 7 GW today. They now yield approximately 50 TWh of electricity per year, i.e. around 25% of current total RE electricity production and about 8% of the total electricity generated (excluding exports)[1].

Biomass power plants have a special role within the available renewable energy technologies. They employ the same basic principle as fossil fuel power plants: The fuel is burned and the energy released is converted first into steam and then into electricity. In other words, biomass plants are much like "normal" power plants, the only difference being that instead of using natural gas, coal or oil (i.e. biomass stored in the Earth's crust, formed millions of years ago) as fuel, they instead use biomass that has been produced recently or decades before. This means that biomass plants also emit CO_2, but in the exact amounts that have been sequestered from the atmosphere as the biomass has grown in the preceding years; hence, biomass power plants can be classed as carbon-neutral.

Accordingly, these power plants also have the same characteristics as conventional power plants: They are largely location independent, i.e. they can be built near consumption centres; they are available at all times; and they can be controlled to adapt to constantly fluctuating electricity consumption levels.

Therefore If it were possible to base a very substantial part of the *Energiewende/* RE expansion on biomass power plants, the entire electrical infrastructure could remain largely unchanged. Specifically, the far-reaching systemic consequences of the *Energiewende* still to be described in this chapter would mostly not occur or would have only a moderate impact. Essentially the only change would be in substituting the fuel for the power plants.

However, this is not possible. The reason for this is simple: The fuel for biomass power plants—mainly biogas derived from corn, other agricultural products and wood—has to be produced in Germany, and the necessary land is simply not available.

If, for example, 500 TWh of electricity (i.e. about 80% of the gross electricity consumption in Germany) were to be produced from biogas, it would require an acreage of more than 20 mio hectares (100 hectares $= 1$ km^2, 259 hectares $=$ 1 square mile); however, the total agricultural acreage in Germany is only 17 mio hectares.

Table 8.1 Use of arable land in Germany

	Millions of hectares	%
Feedstuffs	9.5	57
Food	4.6	27
Energy crops for biogas	1.3	8
Other	1.3	8
Total	**16.7**	**100**

2013 figures; [2]

Approximately 1.3 mio hectares of arable land is required to produce the current level of around 30 TWh of electricity from biogas, which is nearly 8% of the available land in Germany. A further 5% is used for biofuels (see Table 8.1). Increasing this would be difficult, and consequently the "Lead Study 2011" plans only a very modest additional increase in electricity production from biomass from now onwards.

The same also largely applies to solid biomass, i.e. wood. Even now about 50% of timber production in Germany is already used for energy, where use for space heating dominates over use for electricity production.

Of course, one could ask whether it might not be feasible to import biomass fuels. But, firstly, that would thwart the *Energiewende* motive to reduce dependence on energy imports, and, secondly, such an approach would not be economically viable. In any case, biomass has been by far the most expensive RE technology for several years.

Conclusion Biomass has played a significant role in the progress of the *Energiewende so far* and it has attractive properties, but when it comes to the *future of the Energiewende* in Germany, it cannot provide a major contribution.

> From a purely conceptual perspective, one can establish that Germany essentially depends on wind and solar power, i.e. on wind turbines (onshore and offshore) and PV plants, for the expansion of renewable energies in electricity generation.

This simple statement has far-reaching consequences, namely, the systemic consequences of the *Energiewende* that will be illustrated in the following sections.

8.2 Types of Renewable Energy 2

So far in the *Energiewende*, solar and wind power have clearly dominated the expansion of RE. Looking at the renewable energy plants built between 2000 and 2015, wind (onshore) and solar account for more than 90% of the power output and almost 70% of the electricity produced. Offshore wind power, by contrast, has not yet played a major role. We have also just seen that when it comes to the future of renewable energies in Germany, from today's perspective, only solar power and wind power are still available.

However, what will the mutual relationship between the three renewable energy technologies look like in the future?

- Wind (onshore) = wind (on)
- Wind (offshore) = wind (off)
- PV

Would it be useful, for example, to focus on one or two of the three? Furthermore, would it possibly have made more sense previously in the course of expanding RE in Germany, e.g. to build only wind turbines?

To answer this question, let us consider the essential characteristics of the two previous pillars of the *Energiewende*, PV and wind (on), in more detail:

- Both are available in sufficient quantities. Just 2–3% of the surface area of Germany is sufficient to completely supply Germany with PV electricity; wind (on) requires even less.
- Both are heavily location dependent: "in the big picture", i.e. in terms of geographical location in Germany, and "in the small picture", i.e. in terms of the immediate vicinity (shade, wind cover, etc.).
- Both cannot be controlled in terms of power output over time, i.e. electricity production depends entirely on local weather conditions. In other words, an installed capacity of 1 MW means the actual power output available will vary between 0 and 1 MW.
- Both are also highly volatile: The available power output and thus the electricity produced by the plant can vary greatly within short amount of time—within the space of an hour or in extreme cases a quarter of an hour.
- Both are *currently* comparable in terms of cost (when operated in Germany): Today the total cost per kilowatt-hour of electricity is €0.06–0.08 for PV and likewise € 0.06–0.08 for wind (on).

  (However, this was not always the case. In 2005, for example, PV electricity was still about three times as expensive as wind (on) electricity, and in the future the development of the cost could also be quite different.)
- Finally, both PV and wind plants are pure fixed-cost systems. The investment costs for both systems are high, but once they are built, it costs nothing to produce a kilowatt-hour. This is in contrast to conventional power plants (or biomass power plants) in which the fuel cost (i.e. the cost to produce a single kilowatt-hour at an existing plant) plays a significant role.

Conclusion

In terms of the core technical parameters as well as the key economic parameters, the two dominant renewable technologies at this time, PV and wind (on), are quite similar.

Given this conclusion, we must now repose the original question: Would it possibly have been wiser to rely on only one of these two technologies in the *Energiewende* so far?

The answer would seem to be a clear no. It was, and for the foreseeable future will be, the right decision to implement the *Energiewende* based on *two* pillars. The main reasons for this are as follows:

- In 2005 or 2010, it was not possible, or was only partly possible, to predict how both the technologies and the costs of the two RE types would evolve. Even today it is very difficult to predict the potential technological development prospects for PV and wind power and above all what the costs of these technologies will look like in 2030 or even in 2050. The ability to spread the risk was sufficient reason to work on and use both technologies in Germany.
- Both electricity sources complement each other fairly well, both in *geographic terms*—the best conditions for wind power are mostly found in the north of Germany, for PV in the south—and in *temporal terms*: PV delivers high production rates at noon and in summer, wind power in winter and evenly throughout the day. In other words, the weather-related geographical and temporal variations in electricity production average out much better in a combined PV/wind system than in a PV-only or wind-only system.

However, even if these considerations lead us to a clear conclusion, the issue of what role should be played by offshore wind power remains to be answered. Is it required as a third pillar in the future electricity system?

No, a third pillar is obviously not required in the strict sense. The above figures for PV and wind (on) clearly show that they alone are sufficient to provide the 430 TWh per year of electricity supply that is needed in terms of domestic RE according to the Lead Study 2011 [3] for 2050. Nevertheless, the wind (off) technology plays an important role in this planning: It is projected in [3] that wind (off) will produce around 130 TWh of electricity per year in 2050, i.e. the same volume as wind (on). The potential of wind (off) technology in Germany is also significant: Around 8% of the North Sea area available to Germany will have to be utilized to produce the projected 130 TWh per year.

The primary reason for the consideration of wind (off) technology in the implementation plan of the *Energiewende* is that, despite a number of fundamentally similar characteristics—weather dependency, volatility and a fixed-cost system—it has a significant advantage over PV and wind (on): It is clearly more reliable and constantly available.

For a number of years it looked like this advantage would have to be traded off against significantly higher overall cost per kilowatt-hour. But recent technological advances and drastic cost reductions indicate that wind (off)-costs are actually already in a range similar to wind (on) and PV. This means that (as of today) it indeed makes sense to rely on all three technologies for the forseeable future of the *Energiewende*.

> **Conclusion**
> Due to the restricted availability of hydropower and biomass in Germany, in the future the *Energiewende* will have to rely on both of the technologies that already dominate the renewable energy sector, PV and wind (on), as well as on the wind (off) technology.

We should emphasize that the question concerning the optimum ratio of PV to wind (on) to wind (off) in the future is still open at this point. This issue is, and will have to be, largely dependent on how—certainly also in mutual competition between them—the three technologies evolve in the coming decades, especially on the economical side.

To give a more vivid example: It is perfectly conceivable that a 70:30 ratio of PV to wind will turn out to be optimal; and it is just as possible that a ratio of 30:70 will be the best solution in terms of cost-efficiency. In other words, we might see 100 TWh of PV electricity in 2050, but we might also see 200 TWh of PV electricity (or more). *From a purely technical perspective, both scenarios are clearly feasible.*

8.3   Grid Expansion: The Spatial Dimension

The conclusion we can draw based on the preceding sections is that solar and wind (on) must be two central pillars of the *Energiewende* in Germany. We have already demonstrated that both technologies have one thing in common: Electricity production depends greatly on geographic location, "in the big picture" (in the north or south of Germany) and "in the small picture" (the choice of site in a specific region).

However, wind intensity and solar irradiation available in a location, i.e. the quality of the site for a renewable power plant, logically have no bearing on the proximity of this site to any major centres of electricity consumption. A location that is ideal for generating electricity in terms of wind or solar intensity will generally not be the place with the highest demand for energy.

In the big picture, centres of electricity consumption tend to be located in the south and centre rather than in the north of Germany. In the small picture, they are more often found in urban than in rural areas. The southern German states (Bavaria, Baden-Württemberg, Rhineland-Palatinate and Saarland) have a combined electricity consumption of approximately 220 TWh per year, while the northern states (Lower Saxony, Schleswig-Holstein, Mecklenburg-Western Pomerania, Brandenburg, Saxony-Anhalt, Bremen, Hamburg and Berlin) together consume around 140 TWh per year, i.e. only about 65% as much (Table 8.2).

This creates a conceptual issue with respect to the implementation of the *Energiewende*, which existed, if at all, to only a very limited extent for conventional power plants:

- Either the choice of location for the construction of RE plants is based largely on the criterion of "Where is the electricity most needed?" as was previously the case for conventional power plants
- Or the choice of location is based primarily on the criterion of "Where is the electricity yield highest?"

It is obvious that this alternative is irrelevant with respect to the targets and motives of the *Energiewende*. And if we consider high security of supply to be mandatory, the choice between the two alternatives must be made primarily on the basis of the framework condition "affordability/cost-efficiency".

Hence the More Focused Question Is Is it cheaper in terms of macroeconomic cost to select locations for RE plants according to the criterion of maximum electricity production and then to build additional power grids to connect these plants to geographical centres of electricity consumption? Or is it economically more advantageous to locate RE plants as close as possible to consumption centres, i.e. to accept losses in electricity production per megawatt of installed capacity in return for eliminating the need to construct additional power grids (at least to a substantial extent)?

Table 8.2 Electricity consumption by region in Germany (in TWh) in 2015

Region	(Gross) electricity consumption
North	140
Centre	235
South	220
Total	**595**

Various sources on the electricity consumption of the German states, own calculations

With respect to wind (on), the following facts will be helpful in answering this question:

- The wind speed in southern Germany is on average at least 40% less than in northern Germany, especially in coastal regions. Therefore, the electricity yield of a 1 MW wind plant in the north is on average around 2.5–3 times higher than in the south. If wind electricity costs €0.06–0.07 per kWh in coastal areas, the average cost in the south is more than double. Put another way: Even as little as a 10% drop in wind speed will, all else being equal, increase the price of electricity by at least 35%, i.e. by at least €0.02 per kWh.
- Transporting 1 kWh of electricity from the north to the south of Germany—with new overhead power lines to be built—will cost €0.01–0.015 [4].

These few figures clearly imply that, generally speaking, it is indeed more expedient, because it is significantly cheaper, to produce wind (on) electricity where wind conditions are good—i.e. mainly in northern Germany—and then transport it via newly built power grids to the south, than it is to build wind (on) farms mainly in southern Germany, i.e. close to centres of consumption.

And this is exactly how the expansion of wind (on) energy actually proceeded so far. Currently 70% of wind (on) electricity is produced in northern Germany, with the remaining 30% is generated across the rest of the country [5].

In this context, let us address four further aspects:

1. The figures also show that wind (off) technology cannot be eliminated from Germany's future energy mix solely for the reason that it is very away from the consumption centres. Of course, this is a cost disadvantage. However, if the cost of wind (off) technology itself continues to develop positively, then this disadvantage will be offset quickly.
2. When it comes to PV technology, the relevant figures do not suggest a clear direction. PV electricity in southern Germany is on average 10–20% cheaper than it is in the north. However, since geographical and utilization aspects do point in the same direction for PV, in practice this does not play a similar role.
3. Concerning the relationship between urban and rural regions, without going into detail for reasons of space, we can draw similar conclusions. Expanding the distribution grid is likewise cheaper than accepting major compromises in the choice of locations with respect to optimal electricity production.
4. In the most likely scenario of future RE expansion at this time, we can expect a RE production of around 180 TWh per year in northern Germany in 2030, in contrast to an electricity consumption of only 140 TWh per year in northern Germany. From this consideration alone, it is inevitable to transport large quantities of electricity from the north to the south as part of the *Energiewende*.

> **Conclusion**
>
> In terms of macroeconomic cost, it is clearly more sensible to produce wind (on) electricity mainly in the north of Germany and then transport it to southern Germany than it is to build the required wind turbines mainly in the south near the centres of consumption there (which would of course be possible from a purely technical perspective).
>
> Therefore, and also due to the significant role of wind (off) in the RE expansion, it is an essential systemic consequence of the *Energiewende* that new transmission grids will be needed in Germany to a considerable extent.

These considerations become more complex if—as happened in Germany in the summer of 2015—demands are presented and decisions are made that new transmission grids must be built primarily in the form of *underground* power lines. We will address this topic in the second part of the book (Sect. 14.2).

8.4 Volatility: The Temporal Dimension

PV and wind electricity are now, and will continue to be, the main pillars of the *Energiewende* in Germany, that much is certain. In addition to electricity production being dependent on location as discussed in the previous section, these technologies have one other obvious common feature: Electricity production at these plants is completely dependent on the weather and thus on point in time. Production varies greatly over time and there is no way it can be influenced. For an individual plant—no matter where in Germany it is located—this is obvious: The usable power output fluctuates constantly, often within a day but, in any case, within the space of a week, between 0 and 100% of the installed capacity. However, the following is also true for Germany as a whole: Although PV and wind (on) plants with an *installed capacity* of 80 GW are now in operation in Germany—which is nearly the same power output as all its conventional power plants combined—over the course of a year, the *usable power output* of these RE plants together fluctuates between 0 and around 70% of the installed capacity, i.e. between 0 and 55 GW. The average power output over the entire year is only about 15% of the installed capacity, i.e. currently about 13 GW.

In other words, from 1 GW of installed PV capacity, Germany currently gets no more than about 1 TWh electricity per year; 1 GW of installed wind (on) capacity can produce about 1.7 TWh per year in Germany. Compare this to 7–8 TWh per year in electricity from 1 GW of nuclear or 1 GW of lignite-fired capacity.

These few figures have very far-reaching consequences if—as is the overarching goal of the *Energiewende*—an electricity system is to be built largely on renewable energy, which for Germany essentially means solar and wind power.

8.4.1 Three Consequences

1. It takes a lot of installed capacity to generate relatively small quantities of electricity. Generating 330 TWh per year from PV and wind in 2050 would require at least 160 GW of installed capacity ([3], scenario 2011A). However, the average power output required in Germany is only about 60–65 GW, and the maximum power output required (on cold winter days) is 80–85 GW. This means there will be many hours over the course of a year during which the installed PV and wind plants alone produce too much electricity, i.e. electricity that is not needed by electricity consumers in Germany at that time. How to deal with this?

2. Conversely, despite an enormous installed capacity of 160 GW (for comparison: in 2000, before the *Energiewende*, conventional power plants in Germany provided only around 100 GW installed capacity), there will be many hours over the course of a year during which these plants produce much too little to cover the electricity demand in Germany at that time. How to deal with this?

3. Another significant consequence—one which we will not discuss in detail in this book—is the large, uncontrollable fluctuation in power output within short periods of time. If a windy, sunny day is followed by a windless dusk, power output can change by 50 GW or more in the space of just a few hours. This situation was unthinkable only 10 years ago, and the energy companies' conventional technical control systems are not designed to cope. Solutions must be found to these challenges as well.

    Nevertheless, let us focus on items 1 and 2. In short, the (weather dependent, i.e. uncontrollable) *production of electricity* does not match the *demand for electricity* at all. This is in stark contrast to the former electricity system in Germany. The conventional power plants were built and operated every day (i.e. powered up and shut down) such that the demand for electricity was covered exactly at every point in time.

Preliminary Conclusion

The *Energiewende* is not just about simply replacing certain types of power plants with other types of power plants, but rather involves a much more fundamental redesign of the entire electricity system.

8.4.2 Five Options

So how can this challenge be tackled? In principle, i.e. from today's purely technical-conceptual perspective, there appear to be *five options* for Germany:

1. Switch off RE plants

The easiest approach, of course, would be to simply *shut down* some of the RE plants when excess quantities of electricity are generated by PV and wind, i.e. to "discard" the energy available. When too little electricity is generated by PV and wind, the shortfall can be generated in conventional power plants (although this would require maintaining large parts of today's conventional power plant fleet in the long term).

2. Exchange Electricity with Neighbouring Countries

A second option is to increase the *exchange of electricity with neighbouring countries*. Importing and exporting electricity would at least partly offset any over- and underproduction (corresponding willingness and technical possibilities in the neighbouring countries provided, of course).

3. Control the Electricity Demand

The third option is based on a different approach. So far in the electricity system, the demand from electricity consumers in Germany has always been accepted as a given, and electricity production has been controlled to meet this demand. However, this need not always be the case. There are reasonable ways to also control electricity demand—at least to some extent. Of course, this should be done in a way that does not affect comfort in private homes to a large degree and does not impair industrial performance.

Measures of this type used to control the demand for electricity are commonly referred to as demand-side management (DSM).

4. Store Electricity

The fourth, and in some ways perhaps the most obvious, means of tackling this challenge consists in electricity storage systems. If too much electricity is generated from PV and wind power, it can be stored and then made available again when required, i.e. under conditions of little electricity generation.

This is the principle behind the entire German natural gas industry. Vast underground storage facilities are used to bring the largely constant flow of natural gas from the producing countries Russia, Norway and the Netherlands in line with demand, which varies widely over the course of the year (high gas consumption in winter, low gas consumption in summer).

This obvious solution confronts, however, one fundamental problem insufficiently solved so far: It is technically difficult and it is still expensive to store significant quantities of electricity—which leads us to one of the central themes of the *Energiewende*:

"Can we solve the storage problem at a reasonable cost?" This question has been the subject of much controversial debate in recent years.

5. Install Additional Electricity Consumers/Promote Sector Coupling
Finally, there is a fifth way to solve in particular the problem of temporary overproduction: installing *additional* (i.e. structurally not yet existing) *useful electricity consumption systems*, which comprise variable control systems to allow for activation and deactivation whenever electricity production from RE exceeds the current demands of existing electricity consumers. What might such "additional useful electricity consumption systems" look like? Facilities for generating heat (and thus replacing fossil fuels in space heating), plants for producing hydrogen (as fuel for cars thus replacing fossil fuels in transportation) and so on. This would simultaneously create links between the hitherto largely separate energy sectors of electricity, heat and transportation and the so-called sector coupling.

In conclusion, then, there are five options for solving the core problem of the RE technologies PV and wind, that is to say, their complete weather dependence and thus massive yet uncontrollable temporal fluctuations in power output.

8.4.3 Which Is the Best Option?

First, some important notes:

- Looking at the different options, it is clear that none of the options preclude any other. They are all mutually compatible and can be combined.
- Since they all support the targets and motives of the *Energiewende* in the same way, the choice from among these options must be based on compatibility with the framework conditions, i.e. in particular on the criterion of finding the most cost-efficient solution.
- While options 1 and 4 might be capable of solving the matter of fluctuating renewable electricity production alone, this is not possible with option 2 and realistically not with options 3 and 5 either, although the latter do have significant contributions to make.
- The ongoing global expansion of renewable energies entails that in many research institutions and companies around the world, the issue of electricity storage is being intensely worked on. Due to this, it is generally expected that

there will be very significant progress in storage within the next 10–20 years, on the technological side and especially on the cost side.

- Studies on this subject have consistently shown that up to a RE share of around 50% of the total electricity generated—i.e., according to the road map of the *Energiewende*, up to about 2030—the problem can largely be solved by adopting option 1, and up to that point, it is quite clearly the most cost-efficient option at a macroeconomic level.
- Finally, we can make the following highly qualitative, yet quite reliable, statement: Current estimates suggest that every option comes with a nonlinear cost curve, i.e. the greater impact we require an option to have, the more the (absolute but also the) specific costs will rise.

We could now go into detail regarding the advantages and disadvantages, current cost estimates, etc. for these options. However, this would only ever be a snapshot, since due to technological progress the situation might well be very different in 5 years' time.

We Can Therefore Say, in General Terms Only   It makes sense for Germany:

- To rely primarily on option 1 for the next 10–15 years (since periods of significant over- or underproduction will still be very limited in this period)
- To use this time to develop storage technologies, DSM options, sector coupling technologies and grid interconnections within the EU as far as possible
- And then between 2030 and 2035 to decide, or ideally *let the market decide*, the extent to which each option can or has to be used to safeguard the balance between electricity production and electricity demand as the share of RE in the system increases further (and as conventional power plants are gradually decommissioned due to their age). In particular, it should then emerge
  - How many gigawatts of conventional power plants will be needed by 2040 and 2050
  - How many gigawatts of (additional) exports/imports can be contracted and reliably transported
  - Which DSM measures are economically expedient in the long term
  - To what extent additional electricity consumption facilities for heating and for transportation applications can be installed at reasonable cost and so excess CO_2-free electricity be used in other energy sectors as well
  - How many terawatt-hours of storage capacity are required for the *Energiewende* in the future and what technologies—possibly including storage available in neighbouring countries—are most cost-efficient to this end.

Although it is impossible to confirm today, it is probable that after 2030 a mix of all five options will be the most cost-efficient (while safeguarding security of supply) solution.

8.4.4 What Does this Mean for Germany's Electricity System?

In concrete terms, with regard to the systemic consequences of the *Energiewende*, this means (from today's perspective):

- Despite an expansion in RE plants from around 100 GW (2015) to at least 150 GW in 2030 and at least 180 GW by 2050 (with maximum power demand in Germany currently at about 80 GW), conventional power plants will be still required to a considerable extent: in 2030 around 60–70 GW (of about 90 GW today) and in 2050 probably still between 30 and 50 GW.

    In other words, to a considerable extent, conventional power plants will not be replaced by RE power plants, but the *Energiewende requires two power plant systems*: a conventional power plant system (needed to cover the hours during which too little RE electricity is produced despite enormous installed RE capacity) and a RE power plant system.

- In the years 2030–2050, to a significant extent, electricity storage systems as well as additional, variable-use facilities for converting electricity into heat and transportation energy and/or additional electricity exchange options with neighbouring countries will be required to be able to rationally use the electricity produced that exceeds (conventional) demand during many hours of the year.

- In the future—starting in the next decade, then more intensively from around 2030/2035 on—the *Energiewende* is likely to produce complex interactions between a variety of modules, starting with the volatile electricity production from PV and wind plants: fast powering up and shutting down of conventional power plants; use of various types of electricity storage systems; export and import of substantial quantities of electricity; and use of facilities to control the demand for electricity by starting and stopping industrial production plants, heat pumps, night storage heaters; facilities for hydrogen production and so on.

    It is obvious that this will require sophisticated, highly automated control systems. **Therefore, the increasing digitization of the economy in general will play a central role in the energy economy as well.**

Conclusion
- The characteristic shared by PV and wind plants of temporal variations in electricity production will give rise to a significantly broader energy infrastructure in the course of the *Energiewende*. Instead of *one* fleet of conventional power plants, there will have to be *three* different fleets: RE plants, conventional power plants and storage facilities.
- It will also be necessary to intelligently control these three fleets, along with numerous electricity consumers' devices and also new power lines to neighbouring countries, so that in any weather and in any RE electricity production situation, an economical optimum is provided while respecting the primacy of security of supply.

8.5   Fragmentation of the Energy Landscape

The conventional power plant fleet (nuclear, lignite, hard coal and natural gas), which previously dominated electricity production in Germany before the *Energiewende* by more than 90%, was (and is) of a simple structure. There are only a few hundred power plants with a typical installed capacity of 300–1000 MW (0.3–1 GW) and a typical electricity production of 1–10 TWh per year; these are *large-scale complex industrial plants.* The reasons for this are quickly enumerated: vast economies of scale (a small CHP block costs more than five times as much per installed kW as a large-scale power plant), better physical properties and better cleaning of exhaust gases.

This characteristic also entails that only a few companies in Germany have (had) the required expertise and the necessary capital (typically €0.1–1 billion) to build and operate such plants. Despite its size, a typical conventional power plant has only a small footprint: The power plant itself is only about 0.1 km^2 in size, with ancillary facilities covering a maximum 1 km^2.

The RE plants are fundamentally different in this respect:

Economies of scale
The economy of scale is minimal: One wind turbine costs approximately the same (in terms of cost per installed MW) as 10 or 100 wind turbines in a wind farm. (The economy of scale is significant with regard to *individual* wind turbines, and here technological advances are indeed being made, though it seems difficult to achieve more than 3–4 MW per wind turbine (onshore), from today's perspective.) The same applies to PV modules.

Footprint
Renewable electricity generation demands an immense footprint compared to conventional power plants. A 1000 MW solar power plant would require 30 km^2, i.e. a field 5 km × 6 km in size. A 1000 MW wind power plant would have to extend over an area of more than 100 km^2; however, the fields in between the individual wind turbines can still be used for agricultural purposes. Technically and economically these scales are unobjectionable, but in a densely populated country such as Germany, such dimensions are impossible to realize.

(As we have already seen in Sect. 8.1, this applies to an even greater extent to biogas power plants. A 1000 MW biogas power plant would require acreage of more than 3000 km^2 around the plant.)

What Follows from This?
RE power plants are typically much smaller than conventional power plants, and they can really be small indeed without suffering any major cost-related or techni-cal disadvantages: PV plants and biogas plants just a few hundred kW and wind turbines a few MW. Accordingly, the costs of such plants are typically

€1–50 million. (For even smaller plants—which are widespread in the PV sector in Germany—specific costs then increase significantly.)

In Other Words RE power plants are not large complex industrial facilities; rather this technology allows plants to be assembled in a few days or weeks from standard components, in a variety of sizes and without the need for overly complex expertise.

These completely different characteristics with respect to size, expertise and necessary capital have far-reaching consequences:

– Germany's energy landscape is becoming heavily fragmented due to the fast-growing population of RE plants. In lieu of a few hundred power plants, already thousands of wind farms, some 10,000 biogas plants and more than a million PV systems are installed.
– Construction of RE plants is not limited to a few companies, but can be planned, financed and implemented by many stakeholders.
– There is another completely different reason why this fragmentation and plurality of stakeholders are important: The RE plants built between 2000 and 2014 alone called for €170 billion in investments (see the third part of this book). It would have been very difficult for the incumbent energy industry to raise this sum: In the same period, the energy companies only invested €100 billion in the electricity sector, of which €50 billion was required just for the power grids.

More tangibly: On the financing side alone, the *Energiewende* has only worked so far because the expansion of renewable energies can be split into relatively small parts and the respective necessary capital distributed across many shoulders—many investing stakeholders. (In fact, incumbent energy companies account for just 10–20% of investment in RE plants so far.)

Further Aspects
Let us conclude this section by briefly addressing a number of further aspects in this context.

– As we have seen, in the context of the *Energiewende*, the characteristics of PV and wind are systemically causing electricity generation in Germany to become much more fragmented and require support from many more stakeholders than before.

Historically, this was almost a prerequisite for the *Energiewende* to take its recent, rapid course in Germany—not only because of the major investments needed. For a long time, until the end of the last decade, the established energy companies were sceptical about RE and thus very reluctant with respect to investing in RE plants, despite the good ROIs achievable. Other stakeholders were needed to advance the expansion of renewable energies, especially between 2000 and 2010.
– An important side effect of this development is that it automatically provides a solution to one of the main criticisms of Germany's former energy landscape: the "oligopoly" in German electricity generation. In fact, more than 80% of

conventional power plants were previously owned and operated by just four companies: E.ON, RWE, EnBW and Vattenfall. The extent to which the relevant criticism of these companies or of this state of affairs was justified or not, is a question beyond the scope of this book. However, it is important to note that **the** *Energiewende* **is systemically breaking up the "oligopoly of the Big Four", and for a significant part of the political and social spectrum in Germany, this is an additional motive for the** *Energiewende*—**in addition to and beyond the four motives of the** *Energiewende* **illustrated in this book.**

- The significantly larger footprint of RE plants compared to conventional power plants, as mentioned above, has its own implications that we will discuss in the next section.
- These considerations apply to the currently dominant technologies of PV and wind (on) as well as biomass. Wind (off) technology is more like conventional power plants in this respect. With typical installed capacities of 100–500 MW, investment volumes of €0.1–1 billion and more complex technical issues wind (off) plants can typically only be realized by large companies.

Conclusion
- Before the *Energiewende*, electricity generation was dominated by a few large-scale industrial plants and was therefore in the hands of a few major energy companies.
- Production of electricity from RE (onshore) in Germany is, by contrast, characterized by a very large number of plants of different capacities—the largest ones still being smaller than conventional power plants by a factor of ten—and by a variety of different companies and stakeholders that are capable of building, funding and operating these plants.
- The electricity system of the future is thus, at a technical level, a great deal more fragmented and, at an economic and social level, much more complex than it used to be.

8.6 Footprint, Physical Presence of Renewable Energies

As described in the previous section, RE plants have a much larger footprint than conventional power plants.

The, not unjustified, discussion in recent years in Germany about the immense use of land for biogas plants (currently approximately 5000 mi^2) and the topics involved therein—monocultures, "cornification" of the landscape and energy versus food production—will not be addressed here in more detail, firstly, because we can assume that the status quo has been largely accepted by now. Secondly, and more importantly, for cost reasons alone, biomass plants will (in all probability) not play a significant role in the further expansion of renewable energies in Germany. On the contrary, we might even see a gradual decline in biogas plants in favour of

PV and wind plants after 2025 (i.e. after the end of the respective 20-year GREA subsidization). In other words, the very large current footprint of biogas plants is not really a *systemic consequence* of the *Energiewende*—which can be implemented even without biogas plants—but a (presumably partly transitory) status due to *historical reasons*.

How Can We Assess the Issue of Footprint for the Future (Focusing on PV and Wind)?

Taking the target state as planned for 2050, let us assume that around 60–70 GW of wind turbines will be installed onshore and about 70 GW of PV systems (of which about 50% on roofs, i.e. that do not need additional land). Ground-mounted PV systems will then occupy a maximum area of approx. 1000 km^2. The 60–70 GW of wind turbines in wind farms, each comprising 2–20 turbines, will cover a total area of approximately 2000–3000 km^2, of which the turbines themselves would actually only utilize a maximum 200 km^2, i.e. the remaining areas can continue to be used agriculturally.

Looking at these figures, we need to distinguish between three aspects:

- The actual use of land by RE plants is near negligible (<1500 km^2), approximately 0.4% of Germany.
- The footprint of the RE plants (not including biogas plants) is indeed higher, 3000–4000 km^2, but still accounts only about one percent of Germany's land area. It is therefore not of a magnitude that should constitute a serious obstacle to the *Energiewende*.

  We note in this context that currently 2000–3000 km^2 of land has been claimed for opencast lignite mining, which will no longer be the case in the target state in 2050. **Therefore the land balance between the previous and the new electricity system is largely offset.**
- There is a more critical question in relation to the indirect effects. While large PV plants are less conspicuous (also due to their location along traffic routes), largely uncontroversial and expected to remain so, the 2000–4000 wind farms to be installed in Germany by 2050 will in many places become an integral part of German landscapes that cannot be overlooked, and they will have noticeable impacts on the immediate surroundings. This applies, of course, mainly to northern Germany and in particular its coastal regions.

8.7   Consequences for Conventional Power Plants

We have seen in the preceding sections that the characteristics of the main pillars of the German energy transition—PV and wind—will inevitably create a situation in which, despite the extensive construction of new RE plants in the course of the *Energiewende*, conventional power plants will still be needed to a significant extent in the future.

Table 8.3 Conventional power plant fleet in Germany

	2000	2010	2015 (actual)	2030 (planned)	2050 (planned)
Capacity (GW)	100	100	90	60	35
Electricity production (TWh)	540	510	405	250	80
+ Export (TWh)	0	18	52	(Pending)	(Pending)

[1, 3, 6]

Let us reconsider this in numbers ([3], scenario 2011A, 2030 adapted to the German government's current plans) in Table 8.3.

What do these figures mean?

– Aside from the forced shutdown of around 8 GW of nuclear power in 2011, the size of the conventional power plant fleet has remained virtually unchanged in the last 15 years. Large parts (about 70%) of the current fleet will still be needed until at least 2030.
– The utilization of these power plants has barely decreased. Adjusted for exports (i.e. to cover the electricity demand in Germany), the hours of use did go down: In 2015, these conventional power plants produced 75% of the electricity they delivered in 2000 with an installed capacity of about 90%. However, this effect was largely offset by increased exports of electricity. Per gigawatt of installed conventional capacity, almost the same amount of electricity was produced in 2015 as was in 2000 (about 5 TWh per GW).
– However, this rate will drop (excluding electricity exports) to approximately 4 TWh per GW by 2030, and in the decades that follow, in line with the concept of the *Energiewende*, conventional power plants will increasingly serve merely as backups, i.e. to be started up only when at least 180 GW of installed RE capacity—together with RE electricity imports and storage facilities—are not sufficient to cover domestic demand at that time.

According to Germany's current market rules, power plants earn their money exclusively from the amount of electricity they produce. In light of these figures, the question arises as to whether an annual production of 3 TWh or even just 2 TWh per GW (rather than today's 5 TWh per GW) will be sufficient to operate these power plants profitably in a market economy.

Conceptually speaking there are two basic alternatives with respect to this issue:

– Either conventional power plants will have to earn on average significantly more money per kilowatt-hour in the future than they do today
– Or the market rules must be modified such that conventional power plants are paid not only for producing electricity but also for providing guaranteed power output, in other words, for the backup function they perform in the future electricity system.

The choice between these two options *must* be made if the framework condition "market economy in electricity generation" is to be fulfilled. In particular, it must be done in such a way that even in the coming decades, there will be companies willing to operate or invest in conventional power plants.

Failure to do so, i.e. if conventional power plants are not available in the required capacity the obvious consequence will be that, during weather conditions of little wind and little sunlight, Germany's security of supply could not be maintained.

Conclusion

A systemic consequence of the expansion of RE in conjunction with the framework conditions of "security of supply" and "market economy" is that the market rules for conventional power plants need to be designed such that, even in the long term—i.e. with much less annual hours of use—these power plants can be operated profitably in the market to the extent required.

We can further refine these considerations on the basis of the specific conventional power plant fleet in Germany. In the years 2000–2010, when RE had no significant impact on the electricity market yet, power plants in Germany were essentially ranked in order of merit (= deployment order) as illustrated in Table 8.4.

The two power plant types "highest up" in the merit order, i.e. those with the highest variable costs (= cost of producing a kilowatt-hour at an existing power plant, excluding fixed costs for construction and operation), were natural gas and hard coal.

Due to the way the market operates, the variable costs of these two types of power plants thus also *determined the price on the German power exchange (EEX)*, namely, with an approximate ratio of one-third gas to two-thirds hard coal (based on annual hours of use of around 3000 for gas-fired power plants).

The PV and wind plants have variable costs of 0, which means that they rank *at the bottom* of this merit order, gradually driving out the power plants from the market that rank higher in the merit order. There are fewer and fewer hours during which the power plants high up in the merit order are needed to cover electricity consumption in Germany, i.e. are actually being utilized. **In this way, first the**

Table 8.4 Merit order of conventional power plants in Germany, 2000–2010 (in GW; rounded)

Capacity (GW)	Type of power plant
15	Natural gas (excluding CHP)
25	Hard coal (excluding CHP)
22	Lignite
15	CHP (natural gas, hard coal)
20	Nuclear power
5	Water

CHP = cogeneration, i.e. joint production of electricity and heat; [6]

gas-fired power plants are ousted from the market in Germany (except for CHP plants) followed by the hard coal-fired power plants.

From a systemic point of view, this inevitable effect, i.e. this impact automatically created by the *Energiewende* in conjunction with the current market rules in electricity generation, has two major consequences:

- One consequence has already been mentioned: conventional power plants, in particular the gas- and hard coal-fired power plants, are being utilized less and less, bringing their economic viability into question.

  Since these power plants will still be needed for some hours (albeit for far fewer hours than before) to ensure security of supply, complying with the framework condition "market economy" entails that market rules must be (re) designed to guarantee profitability.

- The second consequence is as follows: As the most expensive power plants in terms of variable cost are gradually ousted from the market, their influence on the price decreases—with the logical consequence that the average electricity price on the EEX drops.

In other words:

The expansion of RE necessarily leads to decreasing electricity prices on the energy exchange and thus—because current market rules are such that all power plants earn their money (only) via these prices—falling profitability for all conventional power plants.

Put more tangibly: It is inevitable that in the course of the *Energiewende*—at least with the market rules as they stand today—operators of conventional power plants will see profits decrease.

8.8 Systemic Consequences: Conclusion

Let us summarize the main findings of this chapter.

If the aim is to achieve the three targets of the *Energiewende*, to satisfy the four underlying motives and to simultaneously comply with the three framework conditions, this will inevitably entail a number of significant systemic consequences for the German electricity system in the future. That is, essential characteristics of Germany's future energy landscape are predetermined:

- In the foreseeable future, Germany will have to rely on **sun power and wind power** for the RE expansion in the course of the *Energiewende*; there will be hundreds of offshore wind parks, thousands of onshore wind farms and an even much greater number of PV systems.

- The power grid will have to be considerably expanded; in particular, there will be large **new transmission power lines**, especially from the north to the south of Germany.
- In addition to the vast number of RE power plants and to the conventional power plants (which will then be online for only relatively a few hours per year), there will be a significant volume of **additional technical infrastructure** to manage the temporal fluctuations in PV and wind (on) electricity: large-scale energy storage and small-scale energy storage in private homes; control elements for managing electricity-consuming equipment and devices in industry, and possibly also in private homes; additional power lines to neighbouring countries to capture synergies with their electricity systems; and new facilities that convert electricity into forms of energy that are then available for space heating and transportation purposes (i.e. capturing synergies between the three energy sectors).
- A **digital infrastructure** is superimposed over this that controls the complex interplay between the various infrastructure components.
- These numerous facilities gradually replacing the relatively few conventional power plants will drastically increase the **visually perceptible physical presence of the electricity infrastructure** and thus the number of citizens directly affected by this infrastructure.
- Due to this fragmentation as well as the high investment needs, the *Energiewende* brings with it a development in electricity generation according to which the relevant infrastructure is no longer built, financed and operated by only a few large energy companies, but by a **variety of different stakeholders**. As a result, the economic interests and stakes in relation to the implementation of the *Energiewende* are likewise distributed across a variety of businesses, citizens and institutions.
- The *Energiewende* leads to an ousting from the market of those conventional power plants that are most expensive in terms of the related variable costs—e.g. gas-fired power plants initially then the hard coal-fired power plants—thus to **decreasing electricity prices on the energy exchange** and consequently to decreasing profitability of conventional power plants overall. Further development of the market rules is needed in the longer term to adequately map the gradually changing function of conventional power plants.

The *Energiewende* inevitably entails that the electricity system in Germany becomes much more infrastructure intensive, complex, fragmented, decentralized, capital intensive, as well as distributed between a much larger number of stakeholders than before.

We can assume today with great certainty that an electricity system like this can be *technically* mastered. The real question that springs to mind, however, is how such an energy economy can be *politically* controlled. For although the above-mentioned contours of the future energy landscape in Germany are largely predetermined, there will always be several technical alternatives within these contours. There will always be a multitude of concrete options for how to shape the ongoing implementation of the *Energiewende* and, in particular, how to organize and regulate a market for electricity infrastructure that (rather than a public authority) is expected to govern future developments as far as possible.

The extent of grid expansion, the size and type of storage facilities, the extent and nature of the remaining conventional power plant fleet, small-scale versus large-scale technologies, decentralized elements versus central elements, and many other aspects—these open questions constitute indeed a vast number of available conceptual/technological alternatives within the *Energiewende*. These options will differ in terms of current and future costs, the degree of visual presence and direct impact on citizens and the degree of autonomy awarded to individual regions versus dependence on larger entities. In short, they will differ with respect to a wide variety of interests among many stakeholders.

Political decisions in this arena will necessarily affect a multitude of interests and inevitably favour some interests over others (or at least that will be the perception). Handling all this consistently within a democratic process over several decades—in such a manner that the requisite underlying social consensus on the *Energiewende* is not jeopardized—is probably the most important challenge facing German society as regards the *Energiewende*.

References

1. AGEB (2016) Stromerzeugung nach Energieträgern 1990–2016. http://www.ag-energiebilanzen.de
2. BdEW (2015) Bio-Erdgas: Fragen, Antworten und Argumente. http://www.bdew.de
3. Leadstudy 2011. Langfristszenarien und Strategien für den Ausbau der erneuerbaren Energien in Deutschland, 29 Mar 2012. http://www.dlr.de/dlr/Portaldata/1/Resources/bilder/portal/portal_2012_1/leitstudie2011_bf.pdf
4. Unnerstall (2016) Faktencheck Energiewende. Springer, Berlin, pp 79/80
5. BdEW (2016) Erneuerbare Energien und das EEG: Zahlen, Daten, Grafiken (2016). http://www.bdew.de
6. Fraunhofer ISE. Energy Charts. https://www.energy-charts.de

Summary from an International Perspective

9

1. At its core, the *Energiewende*—the German energy transition (in the electricity sector)—is a political and, given its long-term design, a social project defined by desired characteristics of the German electricity system in 2050:
   1. No nuclear energy—from a 23% share in electricity generation in 2010
   2. (At least) 80% share in electricity generation—from 17% in 2010
   3. Electricity efficiency = approx. 8 €/kWh—from approx. 4 €/kWh in 2010.
2. This target state in 2050—in particular targets 1 and 2—is an inevitable consequence of the three main underlying motives of the German energy transition:
   I. Drastic reduction of CO_2 emissions
   II. Phase-out of nuclear energy
   III. Reduction of dependence on (foreign) fossil fuels.

   There is a very broad and solid consensus in the German political landscape and in German society on these motives, especially I and II.
3. Motives I and III are essentially shared by most countries all over the world, though with quite different priority on their respective political agendas. Motive II, however, is largely unique to Germany.

   The Paris Climate Agreement of 2015 gave an important global push for motive I and must entail—when taken seriously—fundamental energy transitions in many countries.
4. In addition to the target state in 2050, the *Energiewende* design comprises a number of important milestones, primarily:
   – Shutdown of the last nuclear power plant in 2022
   – RE share of (at least) 50% in electricity generation in 2030.

   While the target state 2050 is conceptually inevitable in light of the underlying motives, this is not true of the milestones: They contain a considerable degree of political arbitrariness. The milestones set out an unnecessarily high speed for the *Energiewende*, which in turn makes it unnecessarily expensive (see Chap. 12).

© Springer-Verlag GmbH Germany 2017
T. Unnerstall, *The German Energy Transition*, DOI 10.1007/978-3-662-54329-0_9

5. A fundamental inconsistency in the German energy transition concept is the almost exclusive focus on the German energy system, even though the CO_2 issue is, by its very nature, a global challenge and the German contribution to global CO_2 emissions is only around 2%.

   Accordingly, a rational climate policy for Germany and for any developed country should focus a considerable part of its attention and financial resources on reducing CO_2 emissions *in other countries*, especially developing countries and emerging economies. It is likely that thereby significant CO_2 reduction potentials can be tapped more cost-efficiently and more quickly. (In return, the speed of energy transition in the own country can be somewhat reduced.)

6. In addition to its immediate targets and underlying motives, Germany's *Energiewende* policy sets out to comply with three basic principles or framework conditions:
   - Safeguard security of supply (as far as possible).
   - Insist on cost-efficiency and ensure affordability of electricity (as far as possible).
   - Preserve market economy in electricity generation (as far as possible).

   While the targets and motives of the *Energiewende* are largely undisputed in German politics and society, the proper application of these principles to the specific implementation of the *Energiewende*—i.e. proper application of the framework conditions in current decisions on energy politics—has repeatedly been the subject of intense political debate and social discussion.

7. These three principles certainly represent framework conditions for the energy policy in many other countries as well. Nevertheless, their weighting (also in relation to each other) and their practical application are likely to lead—in the context of the energy transitions to be expected around the world (see item 3)— to energy policies which will differ considerably from country to country.

8. Taking the targets, motives and framework conditions of the *Energiewende* as a basis, then—consistent implementation provided—essential characteristics of the future electricity system in Germany are already defined:
   - Wind and sun as the primary pillars of electricity generation
   - Large new power lines from generation sites to consumption centres
   - New energy infrastructure (especially storage) and new control systems to handle the high temporal volatility of wind and sun
   - Significantly higher fragmentation, decentralization and diversity of stakeholders than in the previous electricity system.

   On the one hand, these systemic consequences largely depend on the specific circumstances in Germany:
   - Natural conditions for hydropower and biomass
   - Natural conditions for PV and wind power (in particular, hours of use for PV systems and wind plants)
   - Distance of the centres of electricity consumption to the regions with most favourable conditions for PV and wind

- Density of population
- Possibility of synergies with neighbouring countries
- Characteristics of the conventional power plant fleet
- Characteristics of the existing power grid
- And others.

On the other hand, it is probably safe to say that most countries—at any rate, most comparable countries in terms of economic development—will have to face at least some of these systemic consequences to some extent in an energy transition project, regardless of their particular conditions.

9. Within these essential contours of the future electricity system in Germany, many features are not determined yet as of today, i.e. they depend on future developments in technology, costs of technologies, economic opportunities and social preferences:
   - Shares of PV vs. wind (on) vs. wind (off) in electricity generation
   - The extent and technologies of storage facilities
   - The extent and nature of demand-side management (DSM)
   - Market rules for electricity generation, in particular for the remaining fossil fuel-fired power plants
   - The technologies of electricity applications for heating and transportation purposes
   - Mix between decentralized and central infrastructure elements
   - And others.

   With respect to these open questions along the further implementation path of the *Energiewende*, in most cases there will be different conceivable options for political regulation. Due to this fact and due to the diversity of stakeholders in the future energy economy with widely differing interests, the consistent political governance of the *Energiewende* in a democracy is and will continue to be a quite challenging task.

10. The predetermined characteristics of a mostly RE-based ("decarbonized") electricity system and the unanswered questions and unresolved challenges are likely to be of considerable relevance in other countries as well—regardless of whether their individual electricity system includes nuclear energy or not. Although the specific answers to these pending issues will be different in detail from country to country, the similarities and the intense efforts already seen around the globe to develop new energy technologies do present manifold opportunities for political, scientific and economic exchange between the nations of the world on the subject of energy transitions.

Part II

The German Energy Transition: Where Does Germany Stand Today?

The Current Situation

Introduction

10

The second part of this book aims to provide an overview of the status of the *Energiewende* in 2015/2016, i.e. about five years after its official launch and 15 years after the expansion of renewable energies (RE) was significantly accelerated by the introduction of the GREA and the first agreement was concluded to shut down all nuclear power plants.

Therefore, the questions subsequently to be answered are:

- Have the **targets** of the *Energiewende*—more precisely, the milestones to be reached on the way to the target state set out for 2050—actually been achieved?
- Have the underlying political **motives** for the *Energiewende* been satisfied as intended along the path taken so far?
- Have the **framework conditions** underlying German energy policy been complied with so far?
- Where does Germany stand with regard to the **systemic consequences** of the *Energiewende*?

Before we address these four questions consecutively, a number of remarks in advance.

1. According to the concept of this book, we will limit ourselves to answering these questions with regard to the "electricity part" of the *Energiewende*, i.e. we will not address the heating and transportation sectors.

     (The percentage of RE in the heating sector has stagnated at 12–13% since 2011, and the share of renewable energies in the transportation sector has stagnated at 5–6% since 2011.)
2. The *Energiewende* is a political and social project with a timeline of 40 years (2010/2011 to 2050); only 15% of that time has elapsed. Given this fact, i.e. the fact that from an sober perspective Germany is still only in the initial phase of the *Energiewende*, one could raise the following objection against the "interim conclusion" proposed in this second part of the book: At such an early stage of a

© Springer-Verlag GmbH Germany 2017

T. Unnerstall, *The German Energy Transition*, DOI 10.1007/978-3-662-54329-0_10

project, it is not reasonably possible to draw an "interim conclusion", and any such conclusion would be of very limited value.

This objection is definitely worth serious consideration. No matter the detailed answers to the above questions, *judgement of the overall "Energiewende" project on this basis is definitely not possible*—that is, unless an aspect had come to light in the relatively short initial period, leading to the certain conclusion that the project is impossible in terms of technology or economics. However, this is clearly not the case.

3. There are three main reasons why we nevertheless devote considerable space to the question *"Energiewende*—where does Germany stand today?" in this book:
   – An analysis of the implementation of the *Energiewende* up until now is instructive. We will see in the course of this second part that, despite the relatively short duration of the overall project so far, the systemic consequences presented in the first part are already clearly visible and tangible.

   In other words, a look at the current situation in Germany does give a thorough impression of structural challenges which will have to be solved in the next 35 years up to 2050 in terms of the hierarchy of energy politics motives, in terms of ensuring compliance with the framework conditions and in terms of the choice of a specific path from the numerous ways of implementing the *Energiewende.*
   – In the public debate in Germany about the *Energiewende*, the issue of whether the implementation so far has been successful or not is also being discussed widely and controversially—giving rise to, actually unduly, differing attitudes to the *Energiewende* as a whole. This part in particular will therefore strive to list the relevant facts to generate a reliable picture.
   – Finally, as regards the international perspective, a closer look at the implementation path taken so far reveals some important experiences concerning an energy transition process. This part will thus allow us to identify important lessons to be learned in view of projects in other countries similar to the *Energiewende.*

4. We will base our assessment throughout this second part on the year 2015, comparing the actual figures of 2015 with the planned figures set as milestones for 2015 in the Lead Study 2011, scenario 2011A. Where already available upon completion of the book, we will also provide the (preliminary) figures for 2016.

Current Status: Targets

The simplest question in this part is that of the current status regarding the three targets of the *Energiewende*, as described in the first part. It can be answered on the basis of objective data.

11.1 Target 1: Shutdown of Nuclear Power Plants

The successive shutdown of the remaining operational nuclear power plants has a fixed schedule and so far is proceeding as planned (Table 11.1). The last milestone was reached in June 2015 with the shutdown of the nuclear power plant Grafenrheinfeld in Bavaria (actually 6 months earlier than planned).

11.2 Target 2: Expansion of Renewable Energies

The planned milestone for RE expansion has clearly been achieved, indeed even exceeded (Table 11.2).

11.3 Target 3: Increase in Electricity Efficiency

Electricity efficiency (GDP/gross electricity consumption) in 2015 was around 9% higher than in 2010. This improvement is slightly greater than the intended increase rate of 1.6% per year (Table 11.3).

© Springer-Verlag GmbH Germany 2017
T. Unnerstall, *The German Energy Transition*, DOI 10.1007/978-3-662-54329-0_11

Table 11.1 Nuclear power plants in Germany in 2015

	Plan 2015	Actual 2015	2016
Power plant capacity (GW)	11	11	11
Number of power plants	8	8	8

[1, 2]

Table 11.2 Renewable energy sources in Germany in electricity generation in 2015

	Plan 2015	Actual 2015	2016
RE capacity (GW)	88	97	102
Quantity of RE electricity (TWh)	167	187	188

Planned values = [1], scenario 2011A, rounded; [3, 4]

Table 11.3 Electricity efficiency in Germany (= GDP/gross electricity consumption) in 2015

	2010	Plan 2015	Actual 2015	2016
Electricity efficiency (€/kWh)	4.3	+ 8%	+ 9%	+ 11%

GDP at 2010 prices; 2010 = average for the years 2009–2011; [1, 3, 5]

References

1. Leadstudy 2011 (Langfristszenarien und Strategien für den Ausbau der erneuerbaren Energien in Deutschland, 29.03.2012). http://www.dlr.de/dlr/Portaldata/1/Resources/bilder/portal/por tal_2012_1/leitstudie2011_bf.pdf
2. BNA (2016) Kraftwerksliste November 2016 (List of power plants). http://www.bundesnetzagentur.de
3. AGEB (2016) Stromerzeugung nach Energieträgern 1990–2016. http://www.ag-energiebilanzen.de
4. ÜNB (2016) Konzept der ÜNB zur Prognose und Berechnung der EEG-Umlage 2016, 2017. http://www.netztransparenz.de
5. AGEB (2016) Energy consumption in Germany in 2015. http://www.ag-energiebilanzen.de

Current Status: Motives

There are also clear answers—mostly based on unambiguous figures—to the question of the current status regarding the four motives of the *Energiewende*.

12.1 Motive 1: Reduction of (German) CO$_2$ Emissions

This item will be described in detail. Firstly, this is the core motive of the *Energiewende*. Therefore, the development of CO$_2$ emissions in Germany is ultimately the most important yardstick for the success of the German energy transition. In other words, the *development of electricity-related CO$_2$ emissions is the most important yardstick for the electricity transition's success* (i.e. of the "electricity part" of the *Energiewende* discussed in this book).

Secondly, there has been intense public debate about this issue, involving an important misunderstanding.

12.1.1 The Facts

There is no doubt that the current situation differs from the plans, not extensively but noticeably (Table 12.1).

This can be explained by the development of the electricity generation mix not proceeding exactly according to plan (Table 12.2).

This juxtaposition permits three statements:

– The decommissioning of nuclear power plants is proceeding on schedule.
– The expansion of renewable energies is ahead of plan.
– A clear difference between planned and actual figures for 2015 involves the ratio of electricity from coal to electricity from natural gas within the fossil fuel sector: Around 20 TWh of electricity has actually been generated from coal rather than from gas as planned.

© Springer-Verlag GmbH Germany 2017
T. Unnerstall, *The German Energy Transition*, DOI 10.1007/978-3-662-54329-0_12

Table 12.1 CO_2 emissions from electricity generation in Germany (excluding electricity exports) (in mio t)

	Plan 2015	Actual 2015	2016
CO_2 emissions	260	270	260

Planned value = [1], scenario 2011A, adapted to the German Environment Agency's calculation method; electricity exports assumed to stem mainly from hard coal power plants; [2]; 2016 = own calculation

Table 12.2 Electricity generation in Germany 2015 (excluding electricity exports) (in TWh)

	Plan 2015	Actual 2015	2016
Nuclear energy	90	92	85
Coal	200	220	208
Natural gas	95	62	81
RE	170	187	188
Others	30	34	33
Total	**585**	**595**	**595**

Planned values = [1], scenario 2011A, all figures rounded; electricity exports assumed to stem mainly from hard coal power plants; [3]

This has had the impact illustrated in Table 12.1 on CO_2 emissions. According to the plan, CO_2 emissions from generating electricity (excluding electricity exports) in 2015 would have to have been approx. 260 mio t (= about 45 mio t lower than in 2010). In fact, they were approx. 10 mio t (= 4%) higher than planned—mainly due to the 20 TWh more of coal electricity compared to the plan.

12.1.2 The Evaluation

1. What is the reason for the noticeable deviation in the actual development of the German electricity mix?

   It is actually very simple. The deployment of conventional power plants in the market is determined by the so-called merit order, i.e. by the variable costs of each power plant. The lower these variable costs, the more often the plant will be used. This relationship is a direct consequence of the framework condition that (conventional) electricity generation in Germany is organized as a market economy. The variable costs in turn are essentially determined by the three parameters of fuel cost, plant efficiency and cost of CO_2 allowances in the European Trading System. Given the market conditions in recent years, even obsolescent hard coal power plants (with poor efficiency) had significantly more favourable variable costs than modern gas power plants; consequently, they were given preference. This effect arises solely from the ratio of world prices for hard coal to European prices for natural gas and CO_2.

In other words, the increase in electricity produced from RE first ousts electricity from (condensation) gas power plants and only then electricity from hard coal-fired power plants (compare Sect. 8.7).

This was essentially foreseeable. To put it more clearly, the deviation from the plan is actually not an error in implementing the *Energiewende* but rather an error in the plan.

2. So how do we assess this in view of the future? Is there an imminent risk of permanently failing the motive to "reduce CO$_2$ emissions"? The answer is no. The gap between the planned and actual reduction in CO$_2$ emissions will not widen further, simply because from now on further increases in the amounts of renewable electricity will oust hard coal electricity. Approx. 60 TWh of electricity from gas power plants come from *cogeneration plants* (CHP) and are therefore largely produced regardless of the merit order. From today's perspective, the gap could even narrow again because in the course of the political enforcement of cogeneration, the electricity production from gas is expected to increase to 70–80 TWh per year and oust corresponding quantities of hard coal electricity. This effect can already be seen in the figures for 2016.

Conclusion
- The fact that CO$_2$ emissions from electricity generation for domestic use have not decreased as much as planned since 2010 is in no way an indication that the *Energiewende* so far is not working. The reason lies in international price developments, which are independent of the *Energiewende*.
- In particular, there is currently no reason to doubt that the *Energiewende* will fulfil the most important of its underlying motives, to reduce CO$_2$ emissions in the long term.

12.1.3 Final Remarks on This Issue

1. Since the German fossil power plants—especially the hard coal-fired power plants—are increasingly less utilized to serve domestic electricity demand due to increasing quantities of RE electricity, while simultaneously being actually competitive when compared to similar power plants in neighbouring countries, they are increasingly producing electricity for export. Thus, they increasingly oust electricity from foreign power plants. This gives rise to some confusion on the matter. Most official statistics relating to CO$_2$ emissions do not distinguish between CO$_2$ emissions due to domestic demand for electricity and CO$_2$ emissions due to electricity exports, although the latter lead to a similar extent to the reduction of CO$_2$ emissions abroad (Table 12.3).

If one looked only at these statistics—as it is often the case—one would conclude that the *Energiewende* is an utter failure with regard to its central motive, because virtually nothing has changed compared to the year 2010.

Table 12.3 CO_2 emissions from electricity generation, official statistics from the German Environment Agency (in mio t)

	2010	2015
CO_2 emissions	315	312

[2]

Table 12.4 CO_2 emissions from electricity generation, breakdown (in mio t)

	2010	2015
CO_2 emissions (officially)	315	312
Of which from domestic electricity consumption	306	270
Of which from electricity exports/imports	9	42

Electricity exports in 2015 assumed to stem mainly from hard coal power plants; [2]; own calculations

However, this conclusion is of course wrong. As a result, the corresponding figures in this book are stated having been adjusted for exports (see Table 12.4).

In an extreme case, it is possible that the *Energiewende* will be a resounding success in the long term since CO_2 emissions from domestic electricity demand will actually drop significantly and on schedule, yet *nominally* (i.e. in terms of official statistics) the "electricity-related CO_2 emissions" in Germany will not decrease at all for a long time because large parts of the existing fossil fuel-fired plant fleet will produce solely for neighbouring countries, with corresponding CO_2 emissions.

2. The German government's current determination to cling to the CO_2 target for 2020 at any cost, despite these facts, is inadvisable. It is not advisable to take into account the more than 40 million tons of CO_2 emitted in Germany in 2015 and 2016 for the production of electricity exported from Germany at all when assessing whether targets have been achieved.

   More generally, we have to distinguish between controllable and uncontrollable effects and between short-term fluctuations in parameters such as the weather, economy growth, fuel prices, etc. and long-term structural changes.

3. Interestingly, the USA is observing the opposite phenomenon. Due to the very low cost of natural gas (thanks to fracking technology), electricity from gas-fired power plants has ousted electricity from hard coal-fired power plants in recent years, resulting in a major drop in CO_2 emissions from electricity generation in the USA. And just as it would be incorrect to attribute the impact in Germany as described above to the *Energiewende*, it would be incorrect to attribute this drop of CO_2 emissions in the USA to an ambitious American climate policy.

12.2   Motive 2: Phase-out of Nuclear Energy

As stated in Sect. 11.1, the phaseout of nuclear energy is currently progressing on schedule. Another part of this effort is that a systematic search for a final disposal site for nuclear waste has been launched, and the work to dismantle the decommissioned nuclear power plants is also underway.

At this point, we should at least briefly mention one issue that has become apparent in recent years—not necessarily for the first time, but surely more so than previously. Regarding the central matters of dismantling nuclear power plants and the search for and construction of a final disposal site, there are still some very challenging, time-consuming and painful discussions ahead in the following years and decades—in terms of technology and locations as well as cost and especially in terms of *the distribution of costs* between macroeconomic stakeholders.

One thing, however, is clear. Even without the *Energiewende*—for example, if Germany were to keep all its nuclear power plants online in 2010 up to the end of their respective technical lifetime—these discussions would have to be conducted and the respective decisions made. The *Energiewende* is merely bringing the debate forward by one or two decades.

12.3   Motive 3: Reduction of Dependence on Fossil Fuels

Given that in this book we discuss the *Energiewende* with respect to electricity, in this section we will address only the question of whether the reduction in coal and natural gas to produce electricity is on schedule, and in particular, whether the imported fuels for power plants (hard coal and natural gas) have gone down according to plan. Oil plays virtually no role in the German electricity sector.

12.3.1 The Facts

Table 12.5 Electricity generation from fossil power plants in Germany (excluding exports) (in TWh)

	2010	Plan 2015	Actual 2015	2016
Natural gas	90	95	62	81
Hard coal	100	80	65	58
Lignite	145	120	155	150
Fossil fuels	**335**	**295**	**282**	**289**
Imported fossil fuels	**190**	**175**	**127**	**139**

Planned Values = [1], scenario 2011A, all figures rounded; electricity exports assumed to stem mainly from hard coal power plants; [3]

12.3.2 The Evaluation

The reduction of dependence on fossil fuels is slightly ahead of plan; and when focusing on the imported fossil fuels, we can see that the corresponding electricity production is decreasing even significantly faster than planned (Table 12.5)—by approx. 30% compared to 2010.

12.4   Motive 4: Promotion of Innovation/Export Opportunities for Germany's National Economy

On this subject we want to be brief: firstly, because this is not a key issue and secondly, because an objective analysis is extremely difficult here. The question revolves around what Germany's national economy would look like today without the *Energiewende*. The answer undoubtedly depends on a multitude of factors, and it cannot be derived in a methodologically rigorous manner in the context of this book.

We therefore include only three brief notes:

–   In the last years, most experts have been quite critical on this issue. They doubt that the *Energiewende* so far has had a (significant) impact on innovative capabilities or on the volume of exports by Germany's national economy. And their arguments do at least appear reasonable.
–   The following is likely to be undisputed: It would have been wise to spend a greater portion of the significant financial resources and economic efforts expended on the *Energiewende* (see the third part of this book) on research and development in the energy sector rather than on building more RE plants, with the aim of better supporting the motive "promotion of innovation/export opportunities".
–   Recently, however, opinions are coming forward that are more positive as regards the future. Ultramodern wind turbines (onshore and offshore), and in particular the technology for smart controls for an electricity system with significant numbers of RE plants that are expected to be developed and used in the next few years and increasingly in the next decade in Germany, are assumed to have good export prospects.

References

1. Leadstudy (2011) Langfristszenarien und Strategien für den Ausbau der erneuerbaren Energien in Deutschland, 29.03.2012. http://www.dlr.de/dlr/Portaldata/1/Resources/bilder/portal/portal_2012_1/leitstudie2011_bf.pdf
2. UBA (2016) Entwicklung der spezifischen CO_2-Emissionen des deutschen Strommix in den Jahren 1990–2015. http://www.umweltbundesamt.de
3. AGEB (2016) Stromerzeugung nach Energieträgern 1990–2016. http://www.ag-energie bilanzen.de

Current Status: Framework Conditions

This chapter is intended to analyse the extent to which the three most important long-term principles of German energy policy—security of supply, affordability/cost-efficiency and market economy—i.e. the framework conditions of the *Energiewende*, have so far been complied with.

Before dealing with these three framework conditions, in turn we note that in recent years, all three issues have given rise to intense debate in German politics and society. In particular, the matter of the cost of the *Energiewende* has been the focus of many articles and discussions in the media as well as, of course, of many political analyses and decisions. For observers from abroad, this is likewise a central, if not *the* central issue.

13.1 Framework Condition 1: Security of Supply

Security of supply is the easiest issue to address in this chapter. We stated in the first part of the book that it can essentially be addressed by looking at only one objective figure: The statistical average duration of the annual power outages for each electricity consumer.

The fact is that this value has *not increased* in recent years *but decreased* (Table 13.1). In other words, security of supply has so far not been negatively affected by the *Energiewende*.

At this point, we must highlight the fact that due to the fluctuating electricity production from approx. 100 GW of RE plants, it is much more challenging, complex and costly today to maintain security of supply than it was 5 or 10 years ago. Nevertheless, the supply remains secure. Obviously, the network operators (and also the responsible authority, the Federal Grid Agency (BNA)) have learned quickly and well how to tackle and overcome these new challenges.

Of course, some residual risk remains—as always—that the long-term average for outages will be significantly exceeded by an extraordinary event in year X. And one could certainly make an argument that the *Energiewende* has increased this

© Springer-Verlag GmbH Germany 2017
T. Unnerstall, *The German Energy Transition*, DOI 10.1007/978-3-662-54329-0_13

Table 13.1 Security of supply in Germany (duration of average power outages in minutes per year)

	2006	2010	2015
Power outage (min)	21.5	14.9	12.7

[1]

residual risk. Still, there are currently no solid arguments for the claim that security of supply represents a fundamental, uncontrollable problem of the *Energiewende*.

Conclusion
The framework condition of security of supply has so far been fulfilled, and there is little to suggest that this may change in the foreseeable future.

13.2   Framework Condition 2: Affordability/Cost-Efficiency

The introduction to this chapter already pointed to the fact that we now come to a central point in this book. The question of the cost and the affordability of the *Energiewende* has for some time been the most critical question pertaining to the project and has been asked by citizens, businesses, academics, trade unions, representatives of the energy industry and also by international observers.

13.2.1 Preliminary Remarks

Let us start with some general observations.

1. German policymakers have undoubtedly been committing a cardinal sin for years in this respect. The *Energiewende* has been described in terms of targets and motives and painted in a positive light, but hardly anybody has ever mentioned the cost of this immense social project. Even worse, when it came to costs, the predictions were wildly incorrect—namely, far too low.

    This cardinal error has been partly corrected in the last years. "The *Energiewende* cannot be had for free" has been the standard phrase for some time. Yet German politics is still relatively far from an open and transparent handling of this issue.

2. Not only in the communication regarding the *Energiewende* but also in the *Energiewende* policy itself, costs did not play a very important role for a long time (until about 2013). This only changed when the GREA surcharge increased in the course of a few years from a negligible amount to a significant element on electricity bills, causing corresponding reactions by the consumers affected and their professional associations.

Again, the German government has responded and is now certainly focused on cost in its *Energiewende* policy. What began in 2013 with the slogan "We put the brake on electricity prices" now pervades every debate about current *Energiewende* developments and legislative decisions in the form of the mantras "The cost increases due to the *Energiewende* cannot continue!" or "Limiting the costs has top priority!"

However, in this section we need to answer the question as to what consequences the years of neglecting the cost aspects have had so far.

3. The answer to the question of macroeconomic costs will determine to a substantial extent the way in which other countries judge the German energy transition. This in turn feeds back into the motives of the *Energiewende* itself: The German energy transition will have a significant impact on the future of global CO_2 emissions only if it supports the momentum of (in the long-term indispensable) energy transitions in other countries—by proving to be *not only technologically feasible but also to be economically affordable*.

13.2.2 Clarifications

The core of the framework condition in question is to ensure the affordability of energy for each energy customer. In this context it is important to realize that the actual cost of electricity—and thus the affordability of electricity—for each customer depends on *two very different issues*:

(1) It depends on the costs of the electricity system for the society as a whole, i.e. on the macroeconomic costs for the national economy.
(2) It depends on how these costs are distributed within the national economy, i.e. on the question which stakeholders—groups of households, all private households, branches of industry, all companies, etc.—have to pay what, when and how.

Likewise, when focusing on the *Energiewende*, there are:

(1) The issue of the costs of the *Energiewende* to the national economy as a whole
(2) The issue of how to distribute these costs within the national economy.

These are indeed conceptionally independent issues, even though in Germany they are both regulated to a great extent by one and the same law, namely, the German Renewable Energy Act (GREA, in German: EEG).

Let us first turn to issue (2).

Currently, according to the GREA, the costs of the *Energiewende* are paid almost exclusively via electricity bills at a fixed amount per kilowatt-hour consumed (mainly the GREA surcharge; see also Sect. 15.5), i.e. via a financial instrument that has nothing to do with the economic capability of the customers.

This is surprising for a task that concerns society as a whole, as in the case of the *Energiewende*.

We want to point out that the same costs might just as well be defrayed by an "*Energiewende* tax" (e.g. x percent of income tax and y percent of capital gains tax). This different distribution of the costs would certainly lead to a different perception within society and thus to different discussions about the costs for households and businesses, i.e. about "affordability".

Furthermore, even a completely different method of financing is also conceivable. The government could create a special fund fed by a capital levy to at least cover the costs caused by the RE plants installed by 2014. In this case the *Energiewende* would (at least partly) be funded not by private households and businesses but by the greater estates in Germany.

We do *not* want to argue here for or against any of these options. We would simply like to make clear that there are **a number of alternatives concerning the question of how to distribute the macroeconomic costs of the *Energiewende* across the national economy**.

However, the choice between these alternatives is not—or is only partway—a *task for energy policy*; rather, it is a task for social, economic, tax and financial policy. To put it another way: The fact that the *Energiewende* currently imposes a certain burden on a particular private household in Germany is actually not a direct result of the *Energiewende*. Rather, it is the consequence of another, *additional* political decision—a decision that could have been made differently (without otherwise modifying the *Energiewende* in terms of targets, motives and framework conditions in the least) and that could still be changed even now.

We will return to the question of the distribution of costs and their impact in more detail in the third part of the book.

In this chapter, we want to focus on issue (1), because this is indeed the question to be posted exclusively to the *Energiewende* policy and it is also the one in the centre of international attention.

Compliance with the framework condition in this context essentially means complying with the principle of cost-efficiency, i.e. to limit the cost of the *Energiewende* as much as possible—in other words, to the unavoidable amount. The question to be answered here should thus be worded as follows: In the implementation of the *Energiewende* so far, has the principle of cost-efficiency been complied with? Or, more directly:

Taking the approx. €20 billion of macroeconomic costs caused by the *Energiewende* in 2015, are these inevitable costs, or could the *Energiewende* up to the present day have been implemented (with the same status regarding targets and motives but) at a much lower macroeconomic cost?

13.2.3 The Evaluation

A number of substantial methodological challenges emerge when one wants to address this question seriously. Since this is not a quantitative academic treatise but a book aiming to create transparency and offer a reliable overview with respect to the key issues pertaining to the *Energiewende*, we cannot give a precise answer to the above question.

However, we can provide a rough estimate based on three aspects that can be regarded as quite reliable (the following figures refer to the year 2015):

1. In the course of the *Energiewende* so far, the previously much more expensive PV technology had an unnecessarily high share of the RE expansion compared to wind power technology. If one-third of the PV electricity (i.e. approximately 13 TWh) were generated by wind power, the costs would be approximately €2.5 billion lower.
2. In the course of the *Energiewende* so far, investors' returns were on average unnecessarily high (see the third part of the book). Had the GREA subsidies been controlled so that the project returns were on average 1% lower, this would have saved a further €1.2 billion.
3. Finally, the speed of renewable energy expansion between 2010 and 2012 was unnecessarily high. During these 3 years, electricity production from renewables increased by approximately 50 TWh, i.e. approximately 17 TWh per year. If the RE expansion were to continue at this pace, the target 80% share in electricity generation would be reached by 2030 already (and not as planned by 2050). This means the rate of RE expansion, and especially of PV, was twice as high as was actually necessary. And that was expensive. If—through better design of the GREA system—10 TWh of PV electricity had been built in 2013–2014 instead of 2010–2012, that would have saved an additional €1.3 billion.

Conclusion

Had these three aspects been managed differently by the *Energiewende* policy using the GREA regulations, which would certainly have been possible in principle, then of the GREA costs in 2015 of around €20 billion, about 25% could have been saved. *25% of the costs of the Energiewende so far could have been avoided.*

At this point let us look ahead and pull an important figure from the third part of the book. The *Energiewende* up to 2014 has caused macroeconomic costs (i.e. subsidies and subsidy commitments) totaling €400 billion net. This shows that the 25% really translate to the very significant amount of around €100 billon.

> **Conclusion**
>
> The framework condition of "affordability/cost-efficiency" has not been complied with during the implementation of the *Energiewende* so far.
>
> Even a rough estimate shows that the consequences of this failure are quite significant. Cost-aware political management could certainly have saved 20% of the costs to the national economy incurred up to 2014 (= €80 billion), in the best case even 25% (= approximately €100 billion).

Now, in such a complex and long-lasting project, we must afford policymakers a certain learning curve as well. Given the conclusion drawn above and the major social and international focus on the costs of the *Energiewende*, it is, however, of crucial importance that the German government actually undertakes the above-mentioned course correction in its future *Energiewende* policy regarding this framework condition, i.e. that it adheres to the maxim that "limiting costs has top priority in the *Energiewende*".

Recent developments (Key Paper published by the German government on the *Energiewende* on 1 July 2015) unfortunately raise doubts in this respect. They rather suggest that in the future, cost-efficiency will still be sacrificed at times on the altar of other political interests and considerations.

13.3 Framework Condition 3: Market Economy in Electricity Generation

Finally, let us take a (much briefer) look at the third framework condition and ask: Is the market economy in electricity generation in Germany still intact, or has it been limited or even overridden by the specific way in which the *Energiewende* is being implemented?

As before it is useful to limit the complexity of this question, which may initially appear deceptively simple. We do so through two preliminary remarks:

1. In a way, this question is essentially already answered in the negative due to the basic design of the *Energiewende*. If the state excludes a key electricity generation technology (nuclear power) from the market and also supports a group of other electricity generation technologies by way of a political imperative with more than €20 billion in subsidies each year (in view of a turnover in electricity generation in Germany of on average around €30 billion per year in 2005–2010), then under no circumstances can we refer to this as a free market economy. We could almost say that the *Energiewende as an energy policy is the downright opposite of a free market economy in electricity generation*. (Economically this can be justified by the fact that significant external (consequential) costs of both nuclear and conventional power plants are not internalized, but such considerations are beyond the scope of this book.)

   However, the question here has a different purpose, namely to discover whether the market economy in electricity generation is limited *beyond this inevitable measure*, i.e. beyond what follows from the concept of *Energiewende* in itself.

2. Recent years have seen major international scientific debate on the following issue. Is the current market mechanism in electricity generation, which has developed mainly in Europe and the USA over the last 15–20 years in essence, prices being established on a national energy exchange and power plants being

deployed according to a "merit order" that defines itself solely based on variable costs—really useful? Are these market rules at all adequate in the long term to meet the essential social needs in the electricity sector, especially the need for security of supply?

This is also an important political issue in Germany and in other countries as well. It is a matter of the "right" market design for the electricity generation market. We will not explore this issue here for two reasons. First, the debate would happen even without the *Energiewende* (as we see from the debate in other countries), though clearly not with the same urgency and stridence. Second, it is a question of the optimal market design *within* the principle of market economy, rather than the question of the extent to which the principle of a market economy itself is upheld.

So, according to these preliminary remarks, let us ignore the fundamental restriction of free competition among electricity generation technologies by the very concept of the *Energiewende* and the debate about the optimum market design for the future and confine ourselves to the question of whether the principles of a free market economy in electricity generation are safeguarded beyond that, i.e. whether in principle any business can build power plants (and shut them down again), whether each power plant can produce electricity (or not) at its own discretion and offer it on the market.

The Answer Is Then:
Largely yes, but with notable restrictions that are, however, exclusively due to the fulfilment of another framework condition, namely security of supply.

In Concrete Terms:
The German government has created two main possibilities to intervene externally in the free play of market forces: the Redispatch Ordinance (*Redispatch-Verordnung*) and the Reserve Power Plant Ordinance (*Reservekraftwerksverordnung*). The Redispatch Ordinance gives TSOs the right to intervene in ongoing power plant operations (i.e. to activate or to shut down power plants) whenever security of supply so requires. The Reserve Power Plant Ordinance permits the Federal Network Agency to prohibit the decommissioning of power plants and to build new power plants outside the market (i.e. in the state-regulated sector) if security of supply so requires.

Explaining these regulations in detail would obviously be a step too far for this book. However, it is clear that:

- Both directives were designed with direct reference to the effects of the *Energiewende*.
- Both are in fact contrary to the principles of a free market economy and therefore should be handled restrictively.

For our purposes, it is important to highlight the extent of these restrictions:

- Redispatch measures were carried out in 2016 totaling approximately 6 TWh [2], in a market of approximately 600 TWh.
- Prohibition of decommissioning pertained to power plants with a capacity of 4–5 GW by the end of 2016 [3], in a market of approximately 90 GW (conventional power plants).

In view of this order of magnitude, it is clear that we cannot speak of a truly significant limitation of the market, at least not yet (although it may be significantly affect individual power plant operators).

Conclusion
The free market in (conventional) electricity generation is still functioning well, by and large, and some consequences—the high annual deployment of lignite- and hard coal-fired power plants, high export volumes and thus nominally relatively high CO_2 emissions in Germany in electricity generation—are noticeable, relevant and (as we have already seen) the subject of much debate.

References

1. https://www.bundesnetzagentur.de/DE/Sachgebiete/ElektrizitaetundGas/Unternehmen_Institutionen/ Versorgungssicherheit/Stromnetze/Versorgungsqualitaet/Versorgungsqualitaet-node.html
2. https://www.netztransparenz.de/EnWG/Redispatch
3. BNA (2016). Kraftwerksliste November 2016 (List of power plants). http://www. bundesnetzagentur.de

Current Status: Systemic Consequences

14

In this chapter of the second part of the book, we want to explore the current status with regard to the systemic consequences of the *Energiewende*. We want to describe:

- The extent to which the unavoidable consequences of the *Energiewende* identified in the first part of the book have already become evident and have been recognized, i.e. to what extent the new characteristics/contours of the future energy landscape in Germany are already visible
- How some of the structural options within these systemic consequences have been handled, i.e. what the energy landscape actually looks like within the main contours.

We address this in the same order in which we developed the systemic consequences conceptually in the first part of the book.

14.1 Types of Renewable Energies

In Chap. 11 we already saw that the overall expansion of renewable energies is progressing somewhat faster than expected: Table 14.1.

The picture as regards the individual RE types is shown in Table 14.2.

The figures show that up to the present day, the RE types' shares of the total RE volume have developed largely as expected in the scenario 2011A outlined in the "Lead Study 2011" used as the basis here.

© Springer-Verlag GmbH Germany 2017
T. Unnerstall, *The German Energy Transition*, DOI 10.1007/978-3-662-54329-0_14

Table 14.1 Renewable energies in Germany in electricity generation, 2015

	Plan 2015	Actual 2015	2016
RE capacity (GW)	88	97	102
Quantity of RE electricity (TWh)	167	187	188

[1–3]

Table 14.2 Renewable energy types in electricity generation in 2015, in % of total TWh

	Plan 2015	Actual 2015	2016
PV	18	21	20
Wind (onshore)	38	38	35
Wind (offshore)	5	4	7
Biomass	27	27	27
Water	12	10	11
Total	**100 (= 167 TWh)**	**100 (= 187 TWh)**	**100 (= 188 TWh)**

[1–3]

14.2   Grid Expansion

The status in 2015 with respect to the very important matter of grid expansion—i.e.
in particular the reinforcement and the new construction of power transmission
lines from the north to the south—is more complex.

14.2.1 The Facts

On the One Hand

Development ran as scheduled. Based on the German government's plans with
respect to the expansion of RE of 2010/2011, the Federal Grid Agency prepared a
grid development plan in cooperation with the four transmission system operators
in 2012. This plan specified the grid expansion required over the next ten to
15 years. Roughly speaking, it gave the following estimated figures for the requisite
grid expansion:

– About 3000–4000 km of new power lines to be realized, in particular by means
  of four new large "electricity highways"
– About 3000–4000 km of reinforcement and modernization of existing power
  lines

The required investments have been estimated at €20–25 billion, which
corresponds, assuming an operating life of the grids of 40 years, to annual costs
of around €2 billion. Add to this the investments needed to connect the grid to the

wind (off) systems (known as the offshore grid development plan), which is expected to require a further €10 billion in investments.

The grid development plan was also fixed in law by the Federal Requirement Plan Act (*Bundesbedarfsplangesetz*) in 2013 so that—at least for the *Energiewende* path up to about 2030—all the necessary foundations have been laid with respect to the grid expansion.

Up to 2013, all this took place on the basis of a very comprehensive consensus of experts, authorities, associations and political parties, as well as the federal government and the state governments.

On the Other Hand

A development occurred in 2014/2015 that tarnished this overall positive picture. In Bavaria massive resistance to the grid expansion plans formed not only among individual concerned citizens and not only within possibly affected municipalities and counties but also at the level of the Bavarian state government. The Bavarian state government revoked the general consensus from 2013 and demanded that the proposed new power transmission lines from northern Germany to Bavaria *not* be built.

14.2.2 The Evaluation

In the overall context of the German energy transition, this turnaround by the Bavarian state government is likely to have been a fleeting (but costly) episode. We will therefore not dwell on it but want to merely point out two key aspects:

- Purely technically, the new transmission lines are in fact *not* necessary, of course. We discussed in the first part of this book that—technically and conceptually—there are a number of very different options for achieving the targets and motives of the *Energiewende* and satisfying the framework condition of security of supply. These options however differ greatly in terms of their respective macroeconomic cost.

  Abandoning the new north-to-south power lines and thus relinquishing the *synergies* between north and south they could realize would generate significant additional costs as well as breach the framework condition of "market economy".
- We must also consider that the real political motive for rejecting the new power lines was that a number of concerned citizens oppose this infrastructure. However, it should be obvious that abandoning the power line infrastructure would entail building other (more expensive) infrastructure in the north and in the south, which would also, naturally, induce local resistance.

These arguments are clear. The extensive political debate in Germany on this topic—sparked mainly by the Bavarian state government—was therefore actually *superfluous*. It did not truly relate to the conceptual, structural options available within the constraints imposed by the targets, motives and framework conditions of the *Energiewende* but ultimately sprang from a lack of political courage to face its inevitable systemic consequences, even if these are contrary to individual stakeholders' interests.

Precisely at this point, it would therefore have been important that the federal government enforce its plans, override the objections of the Bavarian state government and implement its grid expansion plans.

The political reality in Germany is different. A compromise established during summer 2015 does provide for the construction of the two planned power lines to Bavaria, but these will now be primarily realized using underground cabling—a technology that is much more expensive than conventional overhead power lines. According to current, still imprecise, estimates, this will lead to additional investment costs of at least €10 billion.

Thus, yet another time the responsible policymakers have—despite all declarations to the contrary—quite blatantly breached the framework condition of "affordability/cost-efficiency".

> **Conclusion**
> – The systemic consequence "grid expansion" of the *Energiewende* is known, handled and concretized in a comprehensive grid development plan and was put on a solid footing by way of a federal law as of 2013.
> – The political decisions of 2015 on the actual implementation of this grid expansion are expected to lead to the construction of the transmission grids necessary for the expected path of the *Energiewende* (up to about 2030) by 2025. However, they are not compatible with the framework condition of "affordability/cost-efficiency".
> – Recent estimates show that the annual cost of the nationwide grid expansion will increase from less than €1 billion in 2015 to around €5 billion by 2025. This is well below the cost of the RE expansion itself but still represents an appreciable cost factor in the *Energiewende*.

14.3 Volatility

In the first part of the book, we noted that the challenges inherent in weather-dependent electricity production, which fluctuates over the course of weeks, days or even hours (not only at an individual RE plant but also within the entire RE plant fleet in Germany), can basically be mastered by two measures in the next 15 years, i.e., up to around 2030:

– By downregulating RE plants (at times when renewable electricity generation exceeds the demand for electricity in Germany)
– By keeping a large part of the existing conventional power plants online—at least 60 GW, down from 100 GW in 2010 (for times when sufficient renewable

electricity generation is not available, despite an installed RE capacity of well above 100 GW).

In other words, the systemic consequences beyond these two aspects—downregulation of RE, coexistence of RE power plants and conventional power plants—will become fully evident only after 2030.

Considering the situation in 2015 in this respect, we can note three key findings:

1. In 2015, RE plants were downregulated by less than 5 TWh [4]. Given a total RE production of approximately 190 TWh, this is less than 3%.

   Incidentally, most of these downregulations were not required because RE generation in *Germany* exceeded demand in *Germany* but because electricity production in *northern Germany* exceeded demand for electricity in *northern Germany* and, due to the current lack of power lines, the excess could not be transported to the south and its consumption centres.

   In other words, if the grid expansion plans described in the previous section had already been implemented, then today situations necessitating downregulation would be limited to very rare cases.

2. With a capacity of around 90 GW, the conventional power plant fleet in 2015 is only about 10 GW smaller than it was in 2010. This is mainly a result of the politically forced shutting down of around 8 GW of nuclear power in 2011. The specific consequences of this will be discussed in Sect. 14.6.

3. Storage facilities, additional electricity consumers, increased export/import opportunities and DSM measures are not required today and will not be required in the next 15 years either, since even without them, the RE plants can in fact be integrated into the electricity system. Despite this, in Germany there are already a vast number of studies, research undertakings and pilot projects, both regarding individual elements of the future system—power-to-heat and power-to-gas plants, DSM options in industry and private households, large battery storage systems—as well as regarding the possibilities and IT-based implementation of overarching control of these elements.

   To wit, Germany is already undertaking intense preparations for the years following 2030. We can therefore not exclude that development in this area will proceed so quickly, both technically and in terms of cost, that the option to downregulate RE plants and maintain conventional power plants can be replaced in significant parts by the above elements as early as in the next decade.

Be that as it may, from today's perspective there can hardly be reasonable doubt that the challenge of RE volatility will be solved at the *technical level*. And there are also growing indications that this will be possible at costs amounting to only a small proportion of the cost of the RE expansion itself.

One final consideration in this context: There is some evidence that the accumulating technological and engineering expertise in this realm could give rise to interesting export opportunities for Germany's national economy in the foreseeable future (compare Sect. 12.4).

Indeed, the question: "How can a wide variety of decentralized RE plants be efficiently controlled, and the demand for electricity satisfied at every point in time, without an adequate central infrastructure of large-scale power plants and sophisticated grids being available?" is relevant to a whole series of developing countries around the world that are, due to their natural conditions, even more predestined to use RE than Germany.

14.4 Fragmentation of the Energy Landscape

The systemic consequence of the fragmented nature of the RE plant fleet is already impressively visible. In 2015 there were approximately 1.5 million PV plants, 25,000 wind turbines and approximately 14,000 biomass plants spread over Germany.

This *extreme* fragmentation of PV systems, however, was unnecessary. It arose primarily due to the fact that (too) high subsidy rates under the GREA made it financially very attractive for homeowners (especially in the years 2009–2012) to build a PV system on their roof. This has contributed significantly to the breach of the cost-efficiency framework condition described in the previous chapter, i.e. it led to substantial costs of the *Energiewende* that could have been avoided.

14.5 Footprint, Physical Presence of Renewable Energies

Considering the situation in Germany in 2015/2016 with respect to this point, one perceives a mixed picture:

- On the one hand, resistance to new wind turbines is growing rapidly in many places due to noise, the threat to some animal species and optical impairment of the natural landscape.
- On the other hand, 2014, 2015 and 2016 were record years for the expansion of wind power, with about 5, 4 and again 4 GW, respectively, in new wind turbines.

In conclusion, it is probably safe to say that the effects of wind farms on the immediate surroundings have been underestimated for a long time and that more care has to be taken in this respect in the future.

However, from today's perspective, this does not represent a fundamental problem for the further expansion of wind (on) plants in Germany and consequently for the *Energiewende*.

14.6 Consequences for Conventional Power Plants

In this final section on the current status with regard to the systemic consequences of the *Energiewende*, we now come to a subject that has been one of the most complex and contentious issues discussed in German energy politics for quite some time. It concerns the status of the conventional power plants and their future in the next 5 to 10 years, including the current situation and midterm future of the companies that operate these power plants.

14.6.1 Recall

In Sect. 8.7 in the first part of this book, we identified two significant and unavoidable consequences of the *Energiewende*:

- While the conventional power plant fleet will be drastically reduced in the long term, it must remain largely online until around 2030 albeit with greatly reduced hours of use, i.e. significantly less electricity produced per installed capacity. In the longer term, this must lead to a reform of the hitherto applicable market rules.
- In the current market design based on the "merit order", integrating increasing quantities of RE electricity with zero variable costs into the electricity system results in ousting of those power plants that top the merit order (i.e. have the highest variable costs)—with the result that the price of electricity on the energy exchange will drop and thus profitability for all conventional power plants will decrease.

14.6.2 Market Data and Market Rules

If we first look at the market data in electricity generation, we can note:

- The price of electricity on the German energy exchange (EEX) has dropped by €20–25 per MWh in recent years, from an average of €50–55 per MWh in 2005–2010 to around €30 per MWh in 2015.
- There are mainly two reasons for this:

  1. Part of this development is attributable to a marked decrease in the price for hard coal on the global markets in recent years. This decrease has reduced the cost level of hard coal power plants, which are the key factor in defining

prices on the EEX, by around €10 per MWh. This effect is independent of the *Energiewende*.

2. 190 TWh of RE-electricity are now in the system that oust plants higher up in the merit order—notably (condensation) gas power plants—from the market and thus from the price finding mechanism. This is a *direct consequence* of the *Energiewende*.

> Without the *Energiewende*, the EEX electricity price in recent years would have been at least €10 per megawatt-hour (= €0.01 per kWh) higher than was actually the case.

On this basis, the situation for conventional power plants in Germany in 2015/2016 can be briefly summarized as follows:

– Condensation **gas power plants** (with an installed capacity of around 15 GW) are essentially out of the market, i.e. unprofitable. The rational consequence from a microeconomic perspective is the decommissioning of these plants. It is therefore not surprising that around 7 GW of these power plants are already no longer in regular operation or have been registered for decommissioning.
– **Hard coal-fired power plants** run for electricity exports to a considerable extent—indeed, their hours of use to cover domestic demand would be less than 3000 h. In this way, they generally still yield positive operating results but—for the newer plants that have not yet been depreciated—these are insufficient to cover the capital costs.
– **Lignite-fired** and **nuclear power plants,** which in most cases have already been depreciated, are generally profitable but with much smaller profit margins than before.

From a purely market-oriented perspective, this situation will not change a great deal in the next 5 years. Despite a significant deterioration in the business situation for conventional power plants, there will be sufficient installed capacity to guarantee security of supply in Germany.

In the first years of the next decade, however, around 10 GW of nuclear power plants will be shut down within a short period of time. And it is generally expected that, given the prior possible shutdowns of other conventional power plants (in some cases also due to plants reaching the end of their technical lifespan), from 2022 to 2025 on at least some construction of new conventional power plants will be required to guarantee security of supply.

This raises the following question for a successful implementation of the *Energiewende*:

> What do the market rules in electricity generation have to look like for the construction of new conventional power plants to be realized in the market (i.e. by private investors) and for the avoidance of too many shutdowns?

As noted in the introduction to this section, much of the energy policy debate in Germany in recent years has focused on this issue. We do not want to elaborate further on this here but would like to emphasize that, unlike the debate about the power lines from northern Germany to Bavaria, this is a sensible and necessary debate. Indeed, the above question represents an important conceptual issue within the principles of the *Energiewende* as defined by its targets, motives and framework conditions.

The discussion has been brought to an (at least preliminary) end by the new Electricity Market Act adopted in summer 2016. This law does not introduce fundamental reforms (in particular no so-called capacity market), but it does contain substantial further development of the market rules within the current basic market design.

14.6.3 Conclusion

The systemic consequences regarding conventional power plants as derived from the structure of the *Energiewende* in the first part of the book are already visible in 2015:

- Gas-fired power plants (excluding CHP) no longer have a substantial place in the market; they are not required for Germany's electricity supply anymore.
- The price of electricity on the EEX was at least €10 per MWh lower in recent years than it would have been had the *Energiewende* not taken place.
- It might become necessary to construct new conventional power plants—not in this decade, but in the next decade. In any case, substantial decommissioning of power plants (particularly hard coal-fired power plants) should be avoided so as not to jeopardize the security of supply in Germany.
- To this end, revised market rules are required for conventional electricity generation in Germany (at least as long as the framework condition of "market economy in electricity generation" is being upheld). This has been achieved through a new electricity market law effective from summer 2016. Whether this law is sufficient in the longer term to guarantee security of supply is a controversial topic among experts and in the energy industry.
- The operators of conventional power plants—especially RWE, E.ON, EnBW and Vattenfall, but also some municipal utilities—have to cope with massive profit losses in the business area of electricity generation due to the *Energiewende* (in the order of approximately €6 billion per year). This is an inevitable consequence of the *Energiewende* and was essentially to be expected. There is every indication that this situation will not change significantly until at least 2020.

References

1. Leadstudy 2011. (Langfristszenarien und Strategien für den Ausbau der erneuerbaren Energien in Deutschland, 29.03.2012). http://www.dlr.de/dlr/Portaldata/1/Resources/bilder/portal/portal_2012_1/leitstudie2011_bf.pdf
2. AGEB (2016b). Stromerzeugung nach Energieträgern 1990–2016. http://www.ag-energiebilanzen.de
3. ÜNB (2016). Konzept der ÜNB zur Prognose und Berechnung der EEG-Umlage 2016, 2017. http://www.netztransparenz.de
4. EEG in Zahlen 2015. http://www.bundesnetzagentur.de

Summary

15

Looking back at the end of this second part of the book, the answer to the original question as to where Germany stands today—after 5 years of implementation of the *Energiewende*—can be summarized as follows:

1. The **targets** of the *Energiewende*—i.e. the planned milestones for the shutdown of nuclear power plants, the expansion of renewable energies and the increase in electricity efficiency—have been fully achieved and even exceeded.
2. The **motives** of the *Energiewende* have been satisfied accordingly. In particular, CO_2 emissions from electricity generation—for domestic consumption—have fallen significantly, though in 2015 they have been slightly higher than the 2015 target. However:
   - That is no fault of the *Energiewende* but rather a direct consequence of the market economy in conventional electricity generation interacting with global market prices for hard coal and natural gas—and was therefore actually predictable.
   - That does not mean the central motive of a largely CO_2-free ("decarbonized") electricity system in 2050 is at risk.
3. Whether or to what extent the **framework conditions**—security of supply, cost-efficiency and market economy in electricity generation—have been complied with in the implementation of the *Energiewende* so far is the subject of much discussion and controversy. This should come as no surprise since the *Energiewende* inevitably challenges all three framework conditions.

   The *facts*, however, say that the framework conditions of security of supply and market economy have not been significantly limited, beyond the inevitable scope inherent in the *Energiewende* concept; by and large they can be considered as having been complied with.

   By contrast, the facts also say that the framework condition of cost-efficiency has *not* been complied with as yet. 20%, probably as much as 25% of the costs of the RE expansion accrued so far could have been avoided, and these are significant amounts in the region of € 100 billion. The decision to prioritize

© Springer-Verlag GmbH Germany 2017

T. Unnerstall, *The German Energy Transition*, DOI 10.1007/978-3-662-54329-0_15

underground cabling in the nationwide grid expansion will likewise and in addition lead to unnecessarily high costs.

This implies a clear call to policymakers to focus more consistently on the framework condition of cost-efficiency in the future—in actions, not just in words.

4. The **systemic consequences** of the *Energiewende* are, without exception, clearly visible. They are already shaping, to a substantial degree, the reality in the German electricity system. They have also already led to heated debates between and within various stakeholder and policymaker groups. This was—for this is a systemic consequence as well—to be expected in principle. And it is at least in part the outward expression of the inevitable, yet respectable, struggle to find the best way to implement the *Energiewende*.

However, even now—after only 5 years, so actually still in the initial phase of the project—it is evident that the *Energiewende* needs clear political leadership that has the courage to assert its principles against a wide variety of interests.

> Overall, we can say that the *Energiewende* is well on track but has so far been unnecessarily expensive; and it will take wise and steadfast political decision-making to keep it on track at a reasonable cost.

Given this conclusion and also given the paramount importance of the issue of the costs of the *Energiewende* for its prestige in German society as well as for its international reputation, it suggests itself to examine this issue once more in greater detail, i.e. to address the question: "How much does the *Energiewende* cost, and how should we judge these costs?"

The third part of the book is devoted to this question.

Conclusion from an International Perspective   16

Let us sum up the main findings from the second part of the book once more in a different way:

– The *Energiewende* in Germany (in the electricity sector) has so far been successfully implemented.
– This success, however, has come at an unnecessarily high price—Germany has learned the hard way.
– This success is no guarantee that the further implementation of the *Energiewende* in the decades to come will likewise be successful: firstly because the far greater challenges still lie ahead and secondly because it has already become clear in the initial phase that an unusual amount of political determination, political stability and political sense of responsibility is required to successfully manage such a long-term and complex project—and this means above all, to manage it in a cost-efficient manner.
– On the other hand, the chances of a further successful implementation are clearly there from today's perspective. Firstly, renewable energies have become much cheaper in recent years (which is partially also due to the previous expansion in Germany). Secondly, the main cost drivers in the implementation so far have been identified and are being addressed. Thirdly, intensive efforts are underway, and measurable progress is being made at many levels to promptly solve the technical problems predicted to emerge during the next few decades in the context of the *Energiewende*.

From an international perspective, however, it is less important to attempt to predict the further fate of the German energy transition than to discover what lessons can be learned from the implementation of the *Energiewende* in Germany so far. For if the global fight against climate change is to be successful—in terms of the 2 °C goal agreed in Paris—then many other countries will have to "decarbonize" their electricity sectors, i.e. to develop electricity systems based largely on RE (and optionally a larger or smaller proportion of nuclear energy).

© Springer-Verlag GmbH Germany 2017
T. Unnerstall, *The German Energy Transition*, DOI 10.1007/978-3-662-54329-0_16

The chances of the successful implementation of such "energy transitions" increase with the quality of policies and political instruments designed to achieve such an electricity system. One way to improve this quality—at least for similar countries—is obviously to *learn every possible lesson from the German experience.* However, to learn the lessons from the German experience means, above all, to avoid the mistakes made and the weaknesses detectable when we look back on the implementation of the *Energiewende* in Germany in the past 5 years (or 15 years, respectively).

Evaluating the second part of the book once more from this perspective, we believe there are four points worth highlighting in particular:

1. Design of the GREA—expenditure side
2. Design of the GREA—income side
3. Harmonization of grid expansion and RE expansion
4. Communication with/involvement of citizens

Ad 1:
The central political instrument for implementing the *Energiewende* so far has been without doubt the German Renewable Energy Act (GREA; for a description of the basic structures of the GREA, see Sect. 17.5), first introduced in the year 2000. At first glance, the GREA is a good and successful law, and it is no accident that in its basic structure, it has been adopted in over 40 countries. The GREA in its original form is indeed well suited for initial moderate expansion of RE and for gaining experience with integrating renewable energy technologies into a conventional electricity system.

However, when it comes to the systematic, cost-efficient expansion of renewable energies to become the main pillar of the electricity system—i.e. an *Energiewende* in the realest sense—the original GREA contains a crucial weakness.

This weakness consists in the fact that the reimbursements for renewable electricity (the feed-in tariffs in cents per kilowatt-hour produced; for PV, wind and biomass, respectively) are fixed *by the state* in advance for a certain period. However, these reimbursement rates together with market parameters (current prices for RE plants, interest on borrowed capital, etc.) determine the ROI for investors. And if these market parameters change, then ROIs and thus immediately investors' behaviour will also change. While the GREA does provide an option to adjust the reimbursement rates in such a case, this is simply not fast enough process in a parliamentary democracy.

Specifically, in Germany this weakness caused the expansion of renewable energies (especially between 2010 and 2014) to proceed too quickly, with too much focus on PV and with too high ROIs, which altogether generated unnecessary costs in the order of 20–25%.

In other words, it was a weighty and far-reaching mistake to largely keep this basic mechanism of the GREA until 2016. All the more so since this weakness can be relatively easily corrected. In fact, it has been largely resolved in Germany with the reform of the GREA in 2016. The reimbursement rates are no longer defined by the state but *by the market*, i.e. by *tendering* certain volumes (in MW) of RE plants. In this way, firstly the expansion of RE—separately for each type of RE (PV, wind, biomass), if so desired—can be accurately controlled in terms of volume, and secondly, there is a systemic guarantee that only the (minimum) ROI is paid to investors, without which no one would be willing to invest.

A second mistake in the GREA from the viewpoint of cost-efficiency is the high subsidization even for very small PV rooftop systems ranging from several kW to 100 kW. Such minute systems are significantly more expensive than large PV systems: According to the GREA, for many years the reimbursement rates for small PV systems were (and still are) €0.05 to 0.10 per kWh higher than for large (open space) systems. Moreover, they require higher specific expenditure for grid management. Finally, for such small systems the tender model described above is not applicable. In a nutshell, such extreme fragmentation of an RE plant fleet is systemically not required and causes considerable unnecessary costs.

Now one might argue that rooftop PV systems avoid additional use of land. But if the 22 GW of small (< 100 kW) rooftop PV systems today were installed by means of open space systems, this would require only an area of about 700 km^2, which—even in a densely populated country like Germany—does not really pose a problem.

Ad 2:
Another fundamental weakness in the GREA does not concern the reimbursement, i.e. is not on the expenditure side, but on the revenue side. Funding the GREA subsidies via consumers' electricity bills (i.e. through apportionment to every kilowatt-hour consumed in € per kWh) is not really a convincing solution to a task that concerns society as a whole, as is the case with the *Energiewende*. In addition, it unnecessarily endangers the affordability of electricity for all customers, since the number of kWh consumed correlates very little with the customer's economic situation. Thus, this design of the GREA tends to lead to social injustice, and it certainly leads to superfluous discussions about any increase in the surcharge (to be witnessed just recently when the GREA surcharge for 2017 was announced).

There are a number of ways to overcome this weakness, the most obvious one being a funding of the RE subsidies by a mechanism based on the income tax and the capital gains tax.

Ad 3:
A fundamental mistake in the implementation—and actually already in the design—of the *Energiewende* in Germany so far is the failure to synchronize RE expansion and grid expansion from the outset.

The fact that right now—bluntly speaking—the RE electricity is there, but the necessary grids are missing, causes a whole host of problems:

- Extensive redispatch by the TSOs, entailing high costs and also impairing the framework condition of marked economy (and leading to complicated discussions and lawsuits concerning the adequate reimbursements for the power plant operators)
- Tensions with neighbouring countries since their electricity systems are sometimes "flooded" with RE electricity from Germany which cannot be transported within the country
- Compromises as to the optimal locations (especially for wind farms) in the future expansion of RE, entailing further complex processes
- And others.

In short, this mistake has also generated unnecessary costs in the billions and will continue to do so in the next 5–10 years.

Assessing this issue with a view to other countries, the extent of the grid expansion required in the course of an energy transition is certainly highly dependent on the natural conditions and the economic structure and will therefore vary from country to country. Nevertheless, most countries probably need to significantly expand their grids (at least in the interest of a cost-efficient approach) to transport renewable electricity from RE plants to consumption centres.

Now it is rather safe to say that in Germany the time required to build new large transmission power lines is above average—the time required for the approval of the authorities, the time required for dealing with local resistance of affected citizens, etc. (Various attempts to accelerate this process have proven to be more or less unsuccessful). But still, in most countries building large new cross-country power lines over hundreds or thousands of kilometres will also take significantly longer than building new RE plants.

It follows that the rate of RE expansion should be adapted to the speed of grid expansion (and not vice versa, as was attempted in vain in Germany). Or in other words, when designing an energy transition with long-term targets, grid expansion and the expansion of renewable energies should be considered and structured together and afforded concerted milestones.

Ad 4:

This aspect will be mentioned only briefly here: firstly because an entire book could be dedicated to this issue alone and secondly because at this point the political and social conditions and individual governments' political ideas—potentially—will diverge widely from one country to another. The third reason is that this is not a question specific to energy policy.

Therefore, we want to note only in general terms that based on Germany's experiences, it can be very helpful to include the issue of communication with and the involvement of citizens from the very outset when designing an energy transition—especially given the inevitable cost burden on the society and given the need for new and visible energy infrastructure.

Part III

The German Energy Transition: What Does It Really Cost?

A Macroeconomic View

Introduction

17

17.1 Significance of the Costs

This third part of the book is devoted to the macroeconomic costs of the *Energiewende*. We have discussed some important conceptual aspects in this context already in the second part. It will be helpful to summarize these considerations here briefly once more:

1. The cost issue is undoubtedly crucial for the *Energiewende*, for two reasons:
   - The cost issue could seriously impair the social consensus in Germany concerning the *Energiewende*, i.e. the *Energiewende* as a whole could be threatened and slowed down.
   - The cost issue is likely to be the primary factor on which the international assessment of the *Energiewende* depends. In particular, we have seen in Chap. 16 that the most important lessons to be learned from the implementation of the *Energiewende* in Germany pertain to (the lack of) cost-efficiency. So, a better understanding of the financial aspects of the *Energiewende*—or, more generally, of energy transitions to renewable energies—should have a positive impact on climate policies in other countries.

2. Regarding the subject of the costs of the *Energiewende*, we must make a clear distinction between the *costs* incurred at the macroeconomic level and the *distribution of these costs* within the national economy, i.e. the costs borne by each of the economy's individual stakeholders.

   The *latter* is obviously of great importance in terms of the acceptance of the *Energiewende*, i.e. for the social consensus, but it is not really an energy policy issue. The *former* is essential for a neutral point of view and in particular for the above-mentioned international assessment of the *Energiewende*.

© Springer-Verlag GmbH Germany 2017
T. Unnerstall, *The German Energy Transition*, DOI 10.1007/978-3-662-54329-0_17

17.2   Three Phases

For this third part, we will divide the overall German energy transition into **three phases**:

– The first phase starts with the launch of the systematic expansion of RE under the GREA and runs up to the first fundamental reform of the GREA in 2014, i.e. the phase from 2000 to 2014.
– The second phase runs from 2015 to 2030.
– The third phase runs from 2030 to 2050.

These phases each provide for a planned expansion of renewable energies by approximately 130–140 TWh (yearly electricity production). Thus, they each represent about one-third of the total planned expansion of RE of about 400 TWh—from around 30 TWh of renewable electricity production in 1999 to 430 TWh of planned renewable electricity production in 2050 ([1], scenario 2011A).

The costs of the first phase are largely determined, i.e. can be reliably analysed. The costs of the second phase can be estimated (based on conditions as they stand today), permitting important qualitative statements. By contrast, we can only guess at the costs of the third phase at this time—as we can only speculate about the state of and the conditions within Germany's national economy from 2030 onward.

17.3   Methodology

To allow for a better appreciation of this part of the book, a few preliminary remarks on methodology:

– We are not attempting to fully analyse or illustrate all the macroeconomic effects of the *Energiewende*. We have limited ourselves to the *direct costs* and *immediate* financial flows arising from the expansion of renewable energies in Germany. The impact on tax revenue, on employment in Germany (and thus possibly on the social systems), etc. will not be considered.
– Similarly, the costs for phasing out nuclear energy (decommissioning and dismantling nuclear power plants, disposal of radioactive waste) have not been assessed. The reason is that these costs would essentially be incurred anyway (i.e. in case of other fundamental decisions in energy policy) and hence cannot be attributed to the *Energiewende*.
– The figures for the first phase do not include the effects of the wind (off) plants already in place by the end of 2014 (about 1 GW). This is conceptually expedient for a number of reasons—wind (off) plays only a very minor role in this first phase—and is simpler methodically.
– The figures given in this third part are mainly derived from data in documents [2, 3] of the Ministry of Economics and data in document [4] of the German

Association of Energy and Water Industries, as well as from data of the Federal Statistical Office. Thus, we have used only publicly available data, which means the figures can be verified by anybody.

– We do not take future inflation into account.
– As throughout the book, the figures given are typically generously rounded so as not to burden the reader with unnecessary details and to illuminate the essential structures as clearly as possible. Therefore, all data are to be understood as having an accuracy of approximately ±5%.

17.4   Note for the Reader

This part of the book is rather "number heavy" and we have attempted to be as concise as possible. For the reader who is primarily interested in the qualitative, conceptual aspects, Chap. 21 should suffice, in which we summarize the main figures and statements provided in this third part.

17.5   Background: Mechanism of the German Renewable Energy Act (GREA)

The analysis in this part of the book is mainly based on the financial flows that are legally defined and regulated by the GREA. Therefore, we will give a brief and simplified presentation of the essential mechanisms of the GREA.

RE plants—in particular PV plants, wind turbines (onshore and offshore) and biomass plants—are *not economically viable* and will probably not become so in the foreseeable future, in the sense that the market value of the electricity from these plants is not sufficient to cover the cost of capital (= depreciation of investment costs, interest on borrowings and equity).

Therefore, these RE plants have been, and are still, built only if the investors receive subsidies. The GREA guarantees these subsidies in the form of a socalled "feed-in tariff"; i.e., the RE plant receives a fixed, non-variable (i.e. not correlated with inflation) amount per kilowatt-hour for the electricity it produces, for a period of 20 years.

Example (Simplified)
If anyone in Germany decided to build a 3 MW wind power plant and commissioned it by the end of 2016, this investor will receive a fixed reimbursement of, e.g., €0.075 for each kilowatt-hour produced by the plant in the next 20 years. Therefore, if this plant produces an average of 5 GWh of electricity per year, for 20 years the investor receives around €375,000 per year, a total of €7.5 million between 2017 and 2036.

Although these are statutory (incentive) payments—i.e. "subsidies"—this money does not come from a government agency but, as per the GREA, from the network operators, ultimately from the transmission system operators (TSOs) who manage the so-called GREA pot. They receive the electricity from the RE plants and reimburse the investors accordingly. For example, the GREA pot for 2017 (with a volume of approximately €30 billion) paid for or subsidized renewable energy plants built in 2006 (in the 11th of 20 years of funding in total) as well as RE plants built in 2016 (in the first of 20 years of funding in total).

Where Does the Money for the GREA Pot Come from?
A minor portion of that money comes from the resale of the RE electricity by the TSOs on the electricity market, which will earn them around €5 billion in 2017. Most of the money—i.e. the difference between the subsidy payments for the electricity from RE plants and the market value of this electricity, also called the "differential costs"—is paid by electricity consumers in Germany. Each year the TSOs—under state supervision—forecast the differential costs for the following calendar year and apportion this sum to the relevant electricity consumption in Germany. From this is derived the "GREA surcharge", i.e. a value in cents per kilowatt-hour that every electricity consumer in Germany has to pay via their electricity bill (with the exception of energy-intensive industries, which are largely exempt from the apportionment for reasons of international competitiveness).

In 2017, the GREA surcharge amounts to €0.0688 per kilowatt-hour. In 2017, a typical German household with a yearly consumption of 3000 kWh thus pays (3000 × €0.0688 =) €206 into the GREA pot, while a typical, non-energy-intensive company with a yearly consumption of 10 mio kWh pays €688,000. Thus, electricity consumers will pay a total of approximately €25 billion into the GREA pot in 2017.

Actually, the GREA pot forms a kind of *parallel budget* to the federal budget. Citizens and businesses contribute (not a GREA tax but instead) a GREA surcharge to this budget, and the investors of the RE plants receive money from this budget (not from a regulatory body, but guaranteed and monitored by the state)—namely the GREA subsidies—to ensure that their investment is microeconomically profitable.

In the following, the differential costs, i.e. the money to be paid by electricity customers into the GREA pot, will be referred to as "macroeconomic costs of the *Energiewende*". These are the costs attributable to the expansion of RE, accruing at the macroeconomic level and to be paid by the economic stakeholders—mainly private households and businesses.

References

1. Leadstudy (2011) Langfristszenarien und Strategien für den Ausbau der erneuerbaren Energien in Deutschland, 29.03.2012. http://www.dlr.de/dlr/Portaldata/1/Resources/bilder/portal/portal_2012_1/leitstudie2011_bf.pdf
2. BMWi (2013) Energie in Deutschland – Trends und Hintergründe zur Energieversorgung. BMWi, Berlin

3. BMWi (2014) Employment from renewable energy in Germany: expansion and operation –
   now and in the future. BMWi, Berlin
4. BdEW (2016) Erneuerbare Energien und das EEG: Zahlen, Daten, Grafiken. http://www.bdew.
   de

Phase 1 of the Energy Transition (2000–2014) 18

18.1 Costs

18.1.1 The Facts

1. In the years 2000–2014, about 80 GW of RE were built that produce on average around 140 TWh of electricity per year.
2. These RE plants required investments of approx. € 170 billion.
3. The macroeconomic costs of these plants, i.e. the subsidies required under the GREA mechanism (the differential costs), can be expected to amount to approx. € 400 billion (net). They are payable in the years 2000–2034.

   The subsidy mechanism of the GREA is not linked to the rate of inflation. That means the real value of the subsidies (the "present value") will be lower, depending on the rate of inflation during the next 20 years.

   Of these € 400 billion, € 100 billion have been paid in the years 2000–2014, while another € 300 billion is to be paid in the years 2015–2034.
4. From these € 400 billion:
   - Approx. € 170 billion go into refunding the investments (i.e. into depreciation).
   - Approx. € 150 billion go into financing the investments (i.e. into interest on borrowed capital and investors' ROIs).

     Based on an average financing structure of 70% debt to 30% equity, and assuming an average interest rate on the debt of 4%, then this € 150 billion consists of € 60 billion for debt interest and about € 90 billion for investors' returns on the invested equity of approximately € 50 billion. The average ROI amounts to 9–10%.
5. In 2015, the main financial flows were as follows (described in a simplified manner; wind (off) not taken into account):

© Springer-Verlag GmbH Germany 2017

T. Unnerstall, *The German Energy Transition*, DOI 10.1007/978-3-662-54329-0_18

The investors in the 80 GW RE plants received subsidy payments of approx. € 25 billion. From these, they had to spend

– € 11.5 billion on financing the investment (8.5 depreciation, 3 interest on borrowed capital)
– € 5 billion on operating costs of the RE plants
– € 3 billion on biomass fuels,

leaving them with approx. € 5.5 billion as investor's ROI. The subsidy payments were funded from the sale of the RE electricity produced on the energy exchange (€ 5 billion) and from the GREA surcharge (€ 20 billion), paid by the electricity consumers.

6. The differential costs of € 20 billion in 2015 for phase 1 of the *Energiewende* were distributed as follows in the national economy (according to the respective electricity consumption of these groups):

– € 7 billion was paid by private households.
– € 6 billion by (the manufacturing) industry.
– € 4 billion by the trade and services sector.
– € 3 billion by other electricity consumers (public sector, transportation, agriculture).

18.1.2 The Evaluation

1. National Economy

Looking at this figure of approx. € 400 billion for the first phase of the *Energiewende* alone, at first such statements as "The *Energiewende* is incredibly expensive" appear understandable.

However, it is important to consider these figures in the context of the German national economy as a whole:

– The subsidies of about € 100 billion paid between 2000 and 2014 accounted for an average share of 0.3% of GNP in this period; currently they amount to around 0.7%.
– In 2015, the actual cost of € 20 billion for phase 1 of the *Energiewende* accounted for around 9% of energy consumers' total energy costs of approximately € 230 billion.

2. Private Households

When we consider the approximately € 7 billion (net) that German private households have to pay currently for phase 1 of the *Energiewende*, the following aspects are essential:

- € 7 billion per year (plus VAT) equates to approx. 0.5% of consumer spending by private households, which currently totals around € 1600 billion per year.
- For comparison, here are several items that go into annual consumer spending:
  - Approx. € 150 billion for leisure and entertainment
  - Approx. € 140 billion for energy (electricity, heat and transportation)
  - Approx. € 23 billion for alcoholic beverages
  - Approx. € 22 billion for tobacco products.
- € 7 billion (plus VAT) means approx. € 200 per year for the average German household. This is roughly equivalent to an increase of € 0.25 per liter in the price of petrol for cars.
- In 2015 alone, the real wages for German households increased by € 40 billion compared to 2014.

Conclusion
Given these figures one certainly cannot claim that the cost of the *Energiewende* for German households is too high or that "the breaking point is reached". Rather for the German households as a whole, the costs of the *Energiewende* have an impact hardly noticeable.

This explicitly does *not* mean that the costs of the *Energiewende* do not actually represent a significant extra burden on a significant number of *individual* households. This point relates to the issue of how the costs are *distributed* in detail to individual (groups of) stakeholders within the national economy, making it—as discussed in the second part of the book—a question primarily not of energy policy but of social policy.

3. Businesses
The current burden on Germany's businesses from phase 1 of the *Energiewende* amounts to approximately € 10 billion per year. Of this, approximately € 6 billion is distributed to (non-energy-intensive) industry, which generally faces international competition, and around € 4 billion to the trade and services sector, which generally does not face international competition. The energy-intensive industries (chemicals, steel, metallurgy, paper, etc.) are largely exempt from the GREA surcharge and therefore not impacted by the *Energiewende* so far.

Considering industry specifically, it is important to note:

- The gross production value (approximately equal to the sales volume) in the manufacturing sector (excluding energy-intensive industries) is currently some € 1500 billion; the *Energiewende* thus burdens the non-energy-intensive industry with costs in the region of 0.4% of the gross production value.
- The staff costs of these companies are roughly 20% of the gross production value. In other words, a single labour agreement of plus 2% in wages entails a similar burden for industry as the current cost of the *Energiewende*.

> **Conclusion**
> Given these figures, in terms of industry in general, one cannot claim that the
> *Energiewende* imposes an undue burden on German companies or that it
> adversely affects the international competitiveness of the companies.

Here again, this is a *general statement* that does *not* apply in each individual
case. Especially for individual companies where energy costs are significantly
higher than the average across all sectors (= 2% of the sales volume) but that
nevertheless do not fall in the "energy-intensive" category and are therefore not
exempt from GREA apportionment, the *Energiewende* might indeed impair the
ability to compete internationally. However, our statement here is that this is the
exception, not the rule.

18.2   Macroeconomic Effects—Trade Balance

Having considered the immediate impact on the main macroeconomic stakeholders
in the first step, in this second step we want to explore the extent to which phase 1 of
the *Energiewende* alters the relationship between the German national economy
and other national economies, i.e. whether it noticeably changes the **trade balance
of the national economy** (more precisely, the balance of payments in goods and
services).

18.2.1 The Facts

– Of the € 170 billion invested in the RE plants built between 2000 and 2014, an
  estimated 20–35% (= € 40–60 billion) have flown abroad, i.e. 20–35% of the
  biomass, PV and wind plants were imported [1]. During that same period,
  German companies (more precisely, companies producing in Germany) in the
  field of RE plant engineering and construction—whose economic development
  can probably mostly be attributed to the *Energiewende*—exported PV and wind
  plants on a similar scale.
     Thus, with respect to the RE plants themselves, the effects of the
  Energiewende on the trade balance are virtually nil or—given the export-import
  volume of Germany of approximately € 1000 billion per year—moderate
  at best.
– Considering the financial flows to investors in RE plants due to phase 1 in 2015,
  there is a net cash flow to investors of approximately € 17 billion according to
  the following calculation: € 25 billion in GREA payments minus biomass fuel
  costs (as these funds remain national) minus current operating costs (these funds
  likewise remain mostly national).

Table 18.1 Electricity mix with and without phase 1 of the *Energiewende* (in TWh

	Without *Energiewende*	With phase 1 of the *Energiewende*
Nuclear energy	160	95
Lignite	155	155
Hard coal	150	85
Natural gas	90	60
RE	30	170
Other	30	30
Total	**615**	**595**

Assessment based on the circumstances in 2014/2015; excluding electricity exports; assuming gross energy consumption in the scenario "without *Energiewende*" to be at the level of 2010; own calculations

According to studies [2], only a maximum 10–15% of investors are from outside Germany. This corresponds to a cash flow out of the national economy of no more than € 2–2.5 billion in 2015.

– Finally, it should be taken into account that the electricity production of 140 TWh of the RE plants built in phase 1 and the higher increase rate in electricity efficiency together prevent energy imports from abroad.

To roughly quantify these effects, it is useful to contrast the scenario without the *Energiewende* against the situation at the end of phase 1 of the *Energiewende* on the basis of the conventional power plant fleet in 2000–2010 (excluding any electricity exports; Table 18.1).

This means the 140 TWh of RE electricity from the RE plants built in phase 1 and the higher electricity efficiency increase rate oust about 65 TWh of nuclear electricity, about 65 TWh of hard coal electricity and about 30 TWh of natural gas electricity.

Thus, given the prices in the last few years, about € 1–1.5 billion in hard coal imports (about 20 mio t of hard coal) and about € 1–1.5 billion in gas imports (approximately 60 TWh of natural gas) are being avoided, that is, € 2–3 billion saved on energy imports (uranium imports are almost negligible).

18.2.2 The Evaluation

From an overarching perspective—RE plants, financial flows to investors, energy imports—it becomes clear that the *Energiewende* in the years 2000–2014 has had, and is still having, no significant impact on the trade balance of Germany's national economy.

(continued)

> Exports and imports of RE plants are roughly balanced, and payments to foreign producers of hard coal and natural gas amounting to € 2–3 billion per year are roughly replaced by financial flows to foreign investors in RE plants.

18.3   Macroeconomic Effects—Internal Redistributions

In the preceding section, we saw that the trade balance of Germany's national economy has remained virtually unaffected by the *Energiewende*: The positive and negative effects by and large offset each other. Within the national economy, however, the *Energiewende* is causing noticable redistributive effects.

18.3.1 The Facts

As a first step, we exclude the funds needed for financing the RE plants (= € 11.5 billion) from our considerations. Then the following major (mostly) internal effects resulted in 2015 from the first phase of the *Energiewende*:

- € -2–3 billion for producers of hard coal, natural gas
- € -5–6 billion for electricity consumers
- € -5–6 billion for operators of conventional power plants
- € +3 billion for biomass producers
- € +5 billion for managers of RE plants, lessors of land, insurance providers, etc.
- € +5.5 billion for RE plant investors' ROI.

18.3.2 The Evaluation

> This list highlights the most important *redistributive effects of phase 1 of the Energiewende*:
>
> - The (foreign-sourced) fossil fuels hard coal and natural gas are being replaced by (domestic) biomass feedstock.
> - Returns from the energy business in the range of € 5–6 billion per year are redistributed—from the operators of conventional power plants to the investors of RE power plants.

In a sense, this is a completely normal process. An established technology (conventional power plants) is replaced by a new one (RE power plants). This is a *structural change in the (energy) economy* in which there are winners and losers as in any structural change. The magnitude of this redistribution is in any case moderate, i.e. it amounts to tenths of a percent of the GDP or the large financial flows within the national economy, respectively.

However, it is remarkable—and therefore nevertheless unusual—that we are not dealing here with an essentially *technology-driven* development (i.e. a development that happens within the economy itself) but rather with a *politically* enforced development.

Especially for the large companies that operate the conventional power plants in Germany—i.e. the losers when it comes to this structural change—this undoubtedly represents a deep cut with sharp declines in profits. And although these companies have admittedly been able to partly compensate for this *"Energiewende* effect" in recent years by increasingly producing electricity for export (especially at the hard coal-fired power plants), this does not change the overall picture significantly.

In the Second Step The question remains as to how in the macroeconomic picture the current € 11.5 billion per year—or from 2000 to 2034, € 230 billion total—should be judged that are required for financing the RE plants built in the first phase of the *Energiewende* (investment costs plus interest on borrowed capital).

The underlying investment of € 170 billion was indeed financed in the years 2000–2014 by investors in the RE plants (equity = an estimated € 50 billion) and banks (borrowed capital = an estimated € 120 billion) and must be repaid. Under the GREA funding mechanism, these costs are borne by the electricity consumers.

We have already seen that overall these investment funds have remained within the national economy. Put another way, essentially they paid for revenues of the PV, wind and biomass plant companies producing in Germany.

So, the key question to ask is probably: What would have happened with this money had there not been an *Energiewende* and, in particular, the GREA? How would the investors and banks have used these funds? Would this money have been invested in different assets? Into securities?

It is probably obvious that this question cannot be answered in a methodologically reliably manner.

At least we can say this: Through the *Energiewende*, the money has been spent on a purpose that fulfils important political and social motives in Germany. Of course, we can also ask whether it might have been possible, or even more appropriate, to fulfil *different* important social motives, such as education, health or homeland security, with the same effort? Such questions, however, are beyond the scope of this book.

Here we can only ask whether it might have been possible, or even more appropriate, to put the same money into *other investments in low-CO$_2$ technologies outside the electricity sector*, e.g. in the heating or transportation sectors?

To Put It More Sharply Taking it for granted that German society wishes to actively pursue climate protection without nuclear power in its own country and is willing to provide significant internal resources to this end – have these funds been used efficiently in this first phase of the *Energiewende*, or would it have been possible to achieve a much greater impact on German CO_2 emissions with the same effort?

We will address this question briefly in the last section of this chapter.

18.4 CO_2 Efficiency

18.4.1 The Facts

The first phase of the *Energiewende* can be expected to generate costs of some € 400 billion, and thereby approx. 60 mio t of CO_2 emissions per year are avoided for 20 to 25 years (service life of the RE plants). The costs of avoiding CO_2 in this first phase of the *Energiewende* thus amount to approx. € 300 per ton.

18.4.2 The Evaluation

On the One Hand   The costs of avoiding CO_2 are clearly very high at € 300 per ton. There are many studies on the potential for avoiding CO_2 in Germany, where this potential is classified according to the respective avoidance costs. Their findings are unanimous: Many possible measures—especially those aimed at increasing energy efficiency in the heating and transportation sectors—are much cheaper, i.e. have significantly lower CO_2 avoidance costs.

For instance, the very broad 2007 BDI study "Costs and Potentials of Greenhouse Gas Avoidance in Germany" [3] came to the conclusion that in Germany, CO_2 emissions of 130–150 mio t per year can be reduced with CO_2 avoidance costs of less than € 20 per t. Many of the measures available are actually economically viable, i.e. generate no costs and require no macroeconomic subsidies. However, the study also shows that tapping of *further* major CO_2 avoidance potentials, especially in the transportation and heating sectors, would then entail very high costs (> € 300 per t of CO_2).

Against this background, the question arises as to why these very favourable opportunities for CO_2 reduction are not being exploited (more than they are). We do not have the space here to address this issue in more detail; therefore, two observations will have to suffice:

– The central problem with this potential is that it involves a multitude of disparate technologies—production technologies in industry, transportation technologies, technologies in the building sector—that need to be applied by a multitude of stakeholders. Creating policy instruments or incentive programs for this is a much more complex and difficult task than supporting the two technologies of PV and wind power through the GREA.
– Nevertheless, German energy policy should pursue this task with far more consistency and attention than in recent years.

On the Other Hand It must be emphasized that energy efficiency measures and low CO_2 technologies in the heating and transportation sectors are *not* sufficient to reach the target state for the *Energiewende* in 2050 (80% overall reduction in CO_2, down to around 200 mio t). It is absolutely necessary that the electricity sector, which alone emitted approximately 310 mio t of CO_2 in 2010, makes a very significant contribution as well.

But if this is the case, if the *Energiewende* in the electricity sector is indispensable to achieve the central political motive to reduce CO_2 emissions, then the question concerning the maximum CO_2 efficiency of investments essentially should not be:

"Would it have been wiser, or would it be wiser in future, to drive the *Energiewende* in the heating and transportation sectors *instead* of in the electricity sector?"

Rather the question must be (or at least must also be):

"How can we implement the *Energiewende* in the electricity sector in a way that is optimally cost-efficient?"

This issue, however, has been discussed already in the second part of this book.

Conclusion

The issue of cost-efficiency in the political implementation of the motive "reduction of CO_2 emissions" *cannot* be solved in Germany by investing only in energy efficiency or mainly in the transportation and heating sectors. CO_2 emissions must be drastically reduced in the electricity sector (as well).

Thus, the cost-efficiency of the *Energiewende* in the electricity sector is indeed of crucial importance.

References

1. DIW-Wochenbericht Nr.41/2010. https://www.diw.de/documents/publikationen/73/diw_01.c. 362402.de/10-41.pdf
2. Weltenergierat-Deutschland e.V: Energie für Deutschland 2015, 2016. http://www.welten ergierat.de.
3. BDI (2007) Kosten und Potenziale der Vermeidung von Treibhausgas-Emissionen in Deutschland – 2007/2009. http://bdi.eu/media/presse/publikationen/Publikation_Treibhausgase missionen_in_Deutschland.pdf

Phase 2 of the Energy Transition (2015–2030)

19

In the second phase of the *Energiewende*, another 130 TWh of RE electricity will come into the German electricity system. By the end of the second phase, around 300 TWh per year, i.e. at least 50% of the electricity produced in Germany (excluding electricity exports), should originate from renewable energies. According to the milestones set by the German government, this target is to be achieved by around 2030. According to the current plans, the expected ratio of the RE types is roughly 30 GW (= 50 TWh) wind (on), 20 GW (= 20 TWh) PV and 15 GW (= 60 TWh) wind (off) of additional construction during these 15 years.

We also need to consider that, during this period up to 2030—depending on the technical service life of the wind and biomass plants and the (technical) possibilities to extend this service life or operate these systems economically without GREA support—a considerable number of RE plants built in the years 2000 to 2010 will be decommissioned. We estimate that 40–60 TWh of "first-generation" RE electricity will be eliminated in this manner, i.e. will have to be replaced. From today's perspective, we assume that this replacement will primarily be achieved through roughly 10 TWh of PV and 40 TWh of wind (on).

19.1 Costs

19.1.1 The Data

Taking these considerations and current/currently predictable average GREA reimbursements (in cents per kWh) as a basis, the following cost estimate results:

1. Average GREA reimbursements per year (for 20 years): 30 TWh PV × € 0.08–0.09 per kWh = € 2.4–2.7 billion, 90 TWh wind (on) × € 0.06–0.07 per kWh = € 5.4–6.3 billion and 60 TWh wind (off) × € 0.07–0.09 per kWh = € 4.2–5.4 billion, i.e. € 12–14 billion overall.

© Springer-Verlag GmbH Germany 2017
T. Unnerstall, *The German Energy Transition*, DOI 10.1007/978-3-662-54329-0_19

Thus, a total of about € 240–280 billion in GREA reimbursements will be required for phase 2, which we expect to break down as (roughly):

-   € 100–120 billion in investment
-   € 70–80 billion in returns (corresponding to an average project ROI of about 6%)
-   € 70–80 billion in ongoing costs for RE plants

2. From this we need to subtract the value of the electricity produced, which at the electricity prices on the EEX seen in the last years amounts to around € 5–6 billion per year. We can extrapolate this to a total of € 100–120 billion.
3. This means:

> The total costs of the second phase of the *Energiewende*—i.e. the required (net) subsidies for the RE plants—can be estimated to be in the range of € 150–200 billion. In other words: Phase 2 will be much cheaper than phase 1.

19.1.2 The Evaluation

-   The (net) subsidies of € 150–200 billion in total—assuming that there will still be GREA-like mechanisms for investors in RE plants in the future—will be incurred between 2015 and 2050, i.e. over the course of 35 years (compare phase 1: 2000–2034). On average, phase 2 will therefore cost € 5–6 billion per year. Assuming an annual growth rate of 1% for GDP, these costs are equivalent to an average share of around 0.15% of GDP.
-   The annual burden on electricity consumers due to phase 2—again assuming that the mechanism for distributing the costs within the national economy remains essentially the same—will grow relatively quickly (due to the high initial reimbursements for wind power) to around € 15 billion per year until 2025 and then fall significantly again after 2030.
-   We can give a relatively reliable estimate as to how the **overall burden** on electricity consumers due to the *Energiewende* (**phase 1 and phase 2**) will develop until 2030.

    The required (net) subsidies and thus the GREA surcharge can be expected to continue rising until around 2023, to approximately € 27–29 billion per year (i.e. € 0.075–0.085 per kilowatt-hour of GREA surcharge), and to decline again from that point.

    For private households, this means a burden from the GREA apportionment of at maximum € 10 billion per year (or € 12 billion including VAT, i.e. € 300 per household).

A burden of no more than approx. € 8 billion per year can be expected on industry.

Conclusion
- The macroeconomic costs of phase 2 of the *Energiewende* up to 2030 can be estimated to amount to between € 150 billion and € 200 billion and thus be significantly lower than the costs of phase 1.
- This means that, in the next decade (until around 2023), the annual burden on electricity consumers due to the GREA surcharge will increase only moderately to € 0.075–0.085 per kWh (from € 0.0688 per kWh today). In the second half of the next decade, the GREA surcharge will decrease again.
- All qualitative statements regarding phase 1 are thus still valid. For private households, generally speaking, this means no undue burden until 2030. And there is no suggestion that we will see a significant deterioration in the competitiveness of German industry due to the *Energiewende* in the foreseeable future.
- This holds still true if we take into account the other, smaller cost effects of the *Energiewende* we have not mentioned here in detail for reasons of space: Additional costs generated by the necessary grid expansion, costs to enhance electricity efficiency, cost savings due to lower prices on the EEX and others.

19.2 Macroeconomic Effects—Trade Balance

We can be brief here. In view of the fact that:

- In phase 2, around € 7–8 billion will be invested in RE plants every year.
- Today companies in the wind turbine industry producing in Germany alone are reporting sales volumes of over € 10 billion per year,

we do not foresee any significant negative effects on the trade balance of Germany's national economy due to phase 2 of the *Energiewende* during the next 15 years in Germany.

If we compare the electricity mix at the end of phase 1 with the electricity mix to predicted, from today's perspective, in 2030 (end of phase 2), the ensuing figures are shown in Table 19.1.

This means the additional 130 TWh of RE electricity will replace (the remaining) 95 TWh of nuclear electricity and 35 TWh of electricity from hard coal. In addition, the production of electricity (to meet domestic demand) should be about 45 TWh lower than in 2014/2015 due to the increase in electricity efficiency.

As a Result, This Means: In the course of the *Energiewende*, **hard coal-fired power plants will no longer be needed for the electricity supply in Germany in 2030**—always provided the framework condition "market economy in electricity

Table 19.1 Electricity mix comparing end of phase 1 against the end of phase 2 (2030), excluding electricity exports (in TWh)

	End of phase 1 (2014/2015)	End of phase 2 (2030), forecast
Nuclear energy	95	0
Lignite	155	140
Hard coal	85	0
Gas (mostly CHP)	60	80
RE	170	300
Other	30	30
Total	**595**	**550**

own calculations

generation" is largely fulfilled, in particular, provided there are no further political interventions to cut lignite conversion to electricity.

This would avoid hard coal imports amounting to roughly 25 mio t per year with a value of around € 1.3–1.9 billion per year—based on prices in the last years.

Finally, considering the probable financial flows to foreign investors, we can assume an increased share of foreign investment in the RE plants of 20–25%. The net cash flow to investors in RE plants built in phase 2 will amount to an average of around € 9 billion per year, and so we estimate a cash flow of approx. € 2 billion per year out of the national economy.

> **Conclusion**
> No significant (negative) effects on the trade balance of Germany's national economy are to be expected due to phase 2 of the *Energiewende*.

19.3 Macroeconomic Effects—Internal Redistributions

We will not discuss these effects in more detail here. It is clear that they will go in a similar direction as in phase 1. Regarding the required € 100–120 billion of investment funds, we can also pose the question here about alternative use. To this end we refer to Sect. 18.3.

19.4 CO$_2$ Efficiency

The specific costs of CO$_2$ avoidance look much better in phase 2 than in phase 1. Firstly, the costs for the additional 130 TWh of RE electricity are significantly lower. Secondly, we have seen (Table 19.1) that coal-based electricity, which is CO$_2$ intensive, will decrease by 100 TWh between 2014/2015 and 2030. Thus, the CO$_2$ emissions avoided will amount to around 75 mio t, and the CO$_2$ avoidance costs—taking into account additional costs to achieve the increase in electricity efficiency—will be in the range of € 100–150 per ton.

19.5 Summary

Reconsidering, at the end of this chapter, phase 2 of the *Energiewende* in the 15 years leading up to 2030, we can establish the following—despite all uncertainties about the future and in full awareness of how difficult it is to predict developments over a such a long period:

1. The macroeconomic costs of this second phase will be substantially lower than those in the first phase, when estimates are based on current costs for RE electricity and—regarding the wind (off) technology—cost degressions that can be predicted today.
2. Assuming that funding mechanisms for the *Energiewende* remain the same in this period, the burden on electricity consumers will increase again due to the GREA apportionment, but compared to today only in the range of plus € 0.01–0.02 per kilowatt-hour (i.e. a maximum 30% extra).
3. With some certainty, there will not be any negative effects on the trade balance of Germany's national economy in this second phase either.
4. In the second phase as well, the operators of conventional power plants are the main "losers" of the structural change in the course of the *Energiewende*. In particular, the profits from the nuclear power plants will be gone, since the shutdown of these plants will be completed in phase 2. Of course, these companies have every possibility to benefit from the positive opportunities as well.
5. The other central motive of the *Energiewende*—to reduce CO_2 emissions—is satisfied to a slightly larger extent than in phase 1. There will be approximately 75 mio t of CO_2 emissions per year less from 2030 onward.

Phase 3 of the Energy Transition (2030–2050)

In the introduction to this third part of the book, we commented that, from the macroeconomic perspective, no reliable statements can be made about phase 3 of the *Energiewende* from 2030 onward. There are *three reasons* for this:

- It is not feasible to give cost estimates for the construction of RE plants over such a long period.
- From 2030 on, the systemic consequences of the temporal volatility of RE electricity production will be fully brought to bear. Storage facilities must be built, exchange with foreign countries expanded, DSM measures realized and additional electricity applications installed on a larger scale, notably for the transportation and heating sectors. The costs and other macroeconomic implications cannot be predicted today.
- Other important parameters that would be required to provide at least a rough estimate from a macroeconomic perspective equally cannot be predicted today, not even in terms of trends: global market prices for hard coal and natural gas, capital costs and many other factors.

However, one thing is clear. Given that the nuclear power plants will have been completely shut down in phase 2, the further expansion of renewable energies in this third phase will come at the full expense of the fossil fuel-fired power plants. This means CO_2 emissions will decline much more rapidly in phase 3 than in phases 1 and 2—by 140–160 mio t according to plan.

© Springer-Verlag GmbH Germany 2017
T. Unnerstall, *The German Energy Transition*, DOI 10.1007/978-3-662-54329-0_20

Summary (Phase 1 and Phase 2)   **21**

Let us summarize the third part of the book and draw a brief conclusion from our consideration of (not all but) key macroeconomic effects of the *Energiewende* up to the year 2030.

– Until 2030—i.e. up to the milestone where around 50% of Germany's electricity is expected to come from RE—about € 300 billion of investment is required in RE plants. From the beginning (2000) of the renewable energy expansion, this is an average of € 10 billion per year.

   By way of comparison:
   • Relative to the total (gross) capital investment in Germany of about € 500 billion per year, these *Energiewende*-related investments account for about 2%.
   • Private financial assets in Germany are in the range of € 5,000 billion. If the RE investments were fully funded by these private financial assets, approximately € 100 billion, i.e. 2%, would be needed (assuming a financing structure of 30% equity to 70% debt).

– By 2030, the investments in RE plants will have given rise to net subsidies or subsidy commitments of about € 600 billion. From this, in addition to the investments, will be paid the interest on borrowed capital, the operating costs of the RE facilities and—as an essential factor—the returns on investment for the investors (which have been rather high until now).
– The bulk of these subsidies (approx. € 500 billion) must be paid during the period from 2010 to 2035, at an average of € 20 billion per year. The peak can be expected in the middle of the next decade at € 25–30 billion.
– This € 600 billion will basically remain in Germany, i.e. does not affect the balance of payments in goods and services of Germany's national economy.

© Springer-Verlag GmbH Germany 2017   129
T. Unnerstall, *The German Energy Transition*, DOI 10.1007/978-3-662-54329-0_21

– However, the *Energiewende* leads to tangible redistribution or changes in
  financial flows *within* the national economy. To put it bluntly, there are winners
  and losers following this profound structural change in the energy economy.

  The main losers are the operators of conventional power plants who experience
  a significant change in their economic situation.

– The burden on the national economy from the construction of RE plants is
  mitigated by the reduction in electricity prices on the EEX by the *Energiewende*,
  which in 2015 accounted for € 5–6 billion. On the other hand, that burden
  is increased by the necessary grid expansion. By 2030, investments of € 60–70
  billion will be needed, which will then lead to additional annual grid costs (grid
  user fees for electricity consumers) in the range of € 5–6 billion per year.

– Taking this and a few smaller cost factors into account, the average overall cost
  of the Energiewende (phase 1 and phase 2) in the next decades will be between
  € 20 billion and € 25 billion per year—and thus, according to the current
  distribution of costs within the national economy, a burden of € 7–9 billion (net)
  per year for private households and € 6–8 billion per year for industry (plus
  about € 7–8 billion per year for TSS, the public sector, transportation and
  agriculture) can be expected. Generally speaking, this burden does not and
  will not lead to significant impairment in the economic opportunities for private
  households or to significant worsening of the German industry's competitive
  ability.

– Progress on satisfying the predominant motive of the *Energiewende*, the reduction of CO_2 emissions (in this case the CO_2 emissions from electricity generation), will be relatively slow until 2030, despite the considerable costs: The
  CO_2 emissions will have decreased by some 130 mio t compared to the
  "pre-*Energiewende*-level", i.e. from 330 mio t (2000) to approx. 200 mio t in
  2030, which equates to a reduction of approx. 40%. This is mainly due to the
  simultaneous phasing out of CO_2-free nuclear power. As a result, the CO_2
  avoidance cost will be relatively high—higher than those of many other conceivable measures notably in the heating and the transportation sectors.

– Nevertheless, there is no reason today to doubt that the overall target—80% less
  CO_2 emissions from electricity generation by 2050—can be achieved; and this
  target is inescapable if Germany wants to decarbonize its economy.

Conclusion from an International Perspective

22

Looking at the *Energiewende* in Germany from an international perspective and with a view to actual or possible energy transitions in other countries, it is apparent that the *Energiewende* embodies a quite particular path in two important respects:

- Compared to most other countries, Germany massively expanded RE—and in particular PV—very early on.
- Among all similar countries, Germany is the only nation to opt for a full, rapid phaseout of nuclear energy.

Therefore—keeping in mind the aim of the book to identify lessons to be learned from the German example—it is interesting to ask what the costs of the German energy transition would be *without* these two particularities. In other words: It might be instructive to compare the actual costs described in Chap. 21 to the (fictional) costs in three different scenarios:

1. What would the *Energiewende* cost if Germany had started it only in 2015?
2. What would the *Energiewende* cost if Germany realized it without the rapid phase-out of nuclear energy?
3. What would the *Energiewende* cost if Germany had started it only in 2015 *and* without the rapid phase-out of nuclear energy?

Indeed, we will see that, from the consideration of these scenarios, a crude but quite reliable estimate of the costs of an energy transition in similar countries (OECD countries) can be derived.

In preparation for this analysis, we will first present the costs of the German energy transition—as it has actually proceeded since 2000 and as it can be predicted up to 2030—in a different way.

© Springer-Verlag GmbH Germany 2017   131
T. Unnerstall, *The German Energy Transition*, DOI 10.1007/978-3-662-54329-0_22

22.1   The Costs of the (Real) Energy Transition

22.1.1 Phase 1 (2000–2014): Learning Phase, Drastic Reduction in the Costs of PV

This phase was characterized mainly by two developments. At a technical and economic level by a sharp decline in the cost of PV electricity (from approximately € 0.50 per kWh in 2000 to less than € 0.10 per kWh today) and at a political and social level by an intensive learning process regarding the optimal design of the central policy instrument, the GREA.

- **(Net) cost of RE electricity = € 0.13–0.14 per kWh**
  On average, the electricity generated by the RE plants built during this first phase—approx. 140 TWh per year = 23% of Germany's electricity consumption—comes at a cost of € 0.16–0.17 per kWh, with a market value for this electricity of € 0.03 per kWh. The net costs to the German national economy are thus € 0.13–0.14 per kWh for 20 years.
- **Costs amount to 0.35–0.4% of GDP for 35 years**
  Over the entire period during which they are incurred (2000–2034), these costs will average 0.35–0.4% of GDP (assuming an average GDP increase in the future of 1% per year).
- **Burden on private households equals 0.25% of consumption spending**
  Since private households in Germany bear about 35% of the costs of the *Energiewende*, and private consumer spending accounts for about 50–60% of the GDP, households will on average (2000–2034) have to pay about 0.25% of their consumption spending for the first phase of the *Energiewende*.

22.1.2 Phase 2 (2015–2030): Controlled Expansion of RE, Cost Degression Mainly for Wind (Offshore)

In this second phase, we will certainly see—due to the fundamental reforms of the GREA—a better controlled and more cost-efficient expansion of RE, as well as more political emphasis on electricity efficiency. In these 15 years, cost degressions of the RE will continue, especially with respect to wind (off); the rate of this development is, of course, very hard to prognosticate. (Very recent auction results suggest that there will be very substantial cost degressions of large PV-systems in the next years).

- **(Net) cost of RE electricity = € 0.04–0.05 per kWh**
  On average, the electricity generated by the RE plants built during this second phase—approx. 180 TWh per year—will probably cost around € 0.07–0.08 per kWh, with a market value for this electricity of (currently) about € 0.03 per kWh. The net cost to the German national economy for this electricity will thus probably be around € 0.04–0.05 per kWh (for 20 years).

– **Estimated costs will amount to approx. 0.2% of GDP for 35 years**

Considering these costs of the RE electricity, and taking into account the cost of starting to replace the RE plants from phase 1 as they reach the end of their technical lifespan, the cost for the necessary grid expansion to transport this RE electricity to the consumption centres, the price-decreasing effect of the *Energiewende* and a few smaller cost factors, one arrives at the following result. Over the entire period during which they are incurred (2015–2050), the total costs will average approx. 0.2% of GDP (assuming an average GDP increase in the future of 1% per year).

– **Burden on private households equals 0.1–0.15% of consumption spending**

If we assume that customer spending will continue to be on the order of 50–60% of GDP, and that the distribution of the *Energiewende* costs within the national economy remains essentially unaltered, the private households will have to spend 0.1–0.15% of their money on phase 2 of the *Energiewende*.

22.1.3 Phase 1 + Phase 2

Taking the *Energiewende* in Germany between 2000 and 2030, i.e. considering phase 1 and phase 2 together, we arrive at the figures described in Chap. 21. The total cost incurred between 2000 and 2050 will amount on average to around 0.4–0.5% of GDP over these 50 years. This is equivalent to a burden on private households on the order of 0.3% of consumer spending.

22.2 The Costs of an Energy Transition in Germany Beginning in 2015

Let us first consider a scenario where Germany had not started the *Energiewende* until 2015. The starting point in electricity generation for this fictional energy transition—i.e. the electricity mix of Germany in 2015 without the GREA of 2000 and without the politically forced shutdown of nuclear power plants in 2011—would then have looked something like this (compare Table 18.1):

Nuclear energy	160 TWh
RE	30 TWh
Fossil fuels (including others)	425 TWh
Total	**615 TWh**

Furthermore, if we define the milestones for 2030 for this energy transition similar to the real *Energiewende* (shutdown of all nuclear power plants until 2030, share of renewable energies 50% in 2030 and electricity efficiency = + 27% compared to 2015 (+ 1.6% per year)), then the target state in 2030 will read like this:

Nuclear energy	0 TWh
RE	280 TWh
Fossil fuels (including others)	280 TWh
Total	**560 TWh**

Assuming a mix of RE in 2030 of roughly 130 TWh from wind (on), 60 TWh from wind (off), 70 TWh from PV and 20 TWh from hydropower, the necessary expansion of 250 TWh from RE would come at a (net) cost of € 0.04–0.05 per kWh (for 20 years). In a rough estimate, the total cost of such an energy transition (including grid expansion and the other direct financial effects) would be **between € 300 billion and € 400 billion** between 2016 and 2050, averaging out at around € 10 billion per year.

Assuming a GDP growth rate of 1%, this equals **0.25–0.3% of GDP** over these 35 years. Accordingly (assuming the same funding mechanism as in the real *Energiewende*), private households in Germany would have to spend around 0.2% of their resources available for consumption spending each year (i.e. about € 80 per year, at 2015 prices) for such an energy transition.

22.3 The Costs of an Energy Transition in Germany Without a Rapid Phase-out of Nuclear Energy

In this scenario, we assume that the nuclear power plants in Germany would be allowed to run for about 50 years; meaning that some of them would be decommissioned in the next decade, and around 10 plants between 2030 and 2040. (This phase-out timeline is not identical to, but close to the one the German government had planned before the Fukushima nuclear accident in 2011). Assuming, then, that this energy transition would have been launched in 2010, it is clear that, up until 2010, the RE expansion would have happened just like it actually did. Taking the principal concepts of the Energiewende as a basis, we stipulate the targets until 2030 to read: share of CO2-free electricity 50% in 2030 and electricity efficiency $= + 1.6\%$ per year; leading to a target state in 2030 which looks like this:

Nuclear energy	100 TWh
RE	175 TWh
Fossil fuels (including others)	275 TWh
Total	**550 TWh**

Consequently, in this scenario only around 70 TWh of RE electricity would have to be added to the electricity mix between 2010 and 2030; and another 40–60 TWh from RE plants going out of operation would have to be replaced. Here, grid expansion would have to occur at a much slower pace and would therefore play only a minor role on the energy transition bill by 2030.

In a rough estimate, this leads to a total cost of such an energy transition up to 2030 of again **between € 300 billion and € 400 billion**, incurred between 2000 and 2050. This equals an average share of **GDP of 0.2–0.25%** over these 50 years.

22.4   The Costs of an Energy Transition in Germany Beginning in 2015 *and* Without a Rapid Phase-out of Nuclear Energy

Finally, let us briefly consider a scenario where Germany had started to actively decarbonize its electricity system only in the year 2015 and would continue to operate its nuclear power plants until they have reached a technical service life of approx. 50 years.

This translates to a transition from an estimated (fictional) electricity mix in 2015 (compare Table 18.1):

Nuclear energy	160 TWh
Renewable energies	30 TWh
Fossil fuels (including others)	425 TWh
Total	**615 TWh**

to a target state in 2030:

Nuclear energy	100 TWh
RE	180 TWh
Fossil fuels (including others)	280 TWh
Total	**560 TWh**

in order to meet the assumed targets of an at least 50% share of CO_2-free electricity and of a 27% increase in electricity efficiency (compare Sect. 22.2).

The necessary expansion of RE plants producing around 150 TWh per year will come at a net cost of 4–5 ct/kWh, and here only moderate grid expansion would be needed until 2030. The costs of this energy transition would be significantly lower than in the scenarios considered before. We can estimate the costs to be no more than **between € 150 billion and € 200 billion** incurred over the years 2015–2050, equivalent to an average of **less than 0.15% of GDP**. All else being equal, the energy transition bill to the average private household would amount to less than 0.1% of consumption spending—i.e. not even € 40 per year (at 2015 prices).

Conclusion
- The particular path that Germany has taken with its *Energiewende*, i.e. in decarbonizing its electricity system, is indeed unusually expensive. Had Germany not taken this path, but were to start decarbonizing its electricity system only now and postpone the nuclear phase-out for about 15 years, then the total cost accumulated until 2030 would be around 75% lower—with the same level of CO_2 emissions and approximately the same imports of fossil fuels in the year 2030.
- At least for Germany, the following holds true: **A successful energy transition in the sense that half of the electricity should be CO_2-free**

(continued)

> **by 2030, starting now, would not be expensive at all**—neither in terms of the macroeconomic cost to the national economy nor with regard to the burden on private households.

This conclusion leads us to ask: Does this apply to other, similar countries as well? In other words, what would an energy transition in the electricity sector cost in a developed economy, if it was launched today?

22.5   What Would an Energy Transition Cost in Country X?

In the last section of this third part of the book, we want to address the question of what an energy transition would cost in a country similar to Germany (and for our purposes here, we look at the OECD countries)—in other words, what proportion of the gross domestic product would the costs roughly amount to.

What do we mean by the term "energy transition" in this context? For our purposes—based on the respective milestone of the *Energiewende* in Germany—we mean a 50% CO_2-free electricity system by 2030, thus achieving a target state in 2030 in which 50% of a country's (gross) electricity consumption comes from nuclear energy and RE. Of course, this target state marks actually only part of the way to an (almost) fully decarbonized electricity system to be achieved by 2050/2060, but for many countries it would mean a decisive step forward.

By the term "costs" we mean—as before in this book—the financial resources needed at the macroeconomic level to fund the energy transition, over and above those needed to maintain the electricity system in place.

Even when specified in this manner, the question posed in this section may at first glance appear nonsensical, or at best too ambitious—too different, so the obvious objection goes, are the economic circumstances and the existing electricity systems, as well as the geographical conditions for the use of RE, in the 35 OECD countries.

However, a closer look at the conclusion of the previous section and the considerations in this chapter so far reveal that—if one makes certain general and simplifying, yet plausible assumptions—the costs of transforming the electricity system in the above sense depend on three parameters only:

- Share of CO_2-free energy sources (nuclear and renewable) in today's electricity system
- Today's electricity efficiency in the country (i.e. GDP/(gross) electricity consumption)
- Average cost of the RE electricity from the RE plants to be built over the course of the proposed energy transition (i.e. over the period 2017–2030).

Thus, we will take the following assumptions as a basis for our assessment. In this section, we will give all figures in US dollar instead of Euro in order to make easier use of OECD data (http://data.oecd.org).

22.5.1 Assumptions

1. We assume that the country's electricity consumption is constant despite moderate economic growth (see assumption 2). This has held true on average for the OECD countries over the last 10 years.
2. We assume an average real GDP growth in the years 2017–2050 of 1% annually. This is a deliberately cautious assumption. The average real GDP growth in the OECD countries was 2% over the past 10 years. (As in general in this third part of the book, we will not take into account inflation here).
3. We assume that the production of electricity from nuclear sources (in case there are any) will not change in the years leading to 2030—i.e. the technical lifespan of the nuclear power plants currently online extends at least until 2030. This assumption entails that (as per assumption 1) the present share of nuclear power in the electricity mix will remain constant and consequently that the energy transition is achieved solely by expanding renewable energies.
4. We assume that the expansion of renewable energies is achieved by means of a funding instrument that, for 20 years, subsidizes the electricity from the RE plants to be built at a fixed reimbursement rate (in $-cents per kWh).
5. The resulting (net) costs of RE electricity (i.e. the necessary direct subsidies less the market value of the RE electricity) are assumed to be $ 0.07 per kilowatt-hour (for 20 years). This is a very cautious assumption at the upper limit of the costs that can actually be expected; it does not yet take into account the major cost degressions of PV- and wind electricity most experts expect over the next 5-10 years. Conversely we assume, somewhat optimistically, that the other financial effects of the envisaged energy transformation—grid expansion costs, cost savings due to falling market prices for electricity, investments avoided in fossil fuel-fired power plants, costs to achieve the increase in electricity efficiency (assumption 1) and others—will roughly offset each other or will at least be much less than the cost for the RE expansion (or will be covered by the assumed $ 0.07 per kilowatt-hour).
6. We assume that in today's electricity system of the country, the current proportion of the CO_2-free energy sources nuclear energy and hydropower (which provide temporally constant electricity) is not less than 15%. This is the case for the OECD countries on average and for most individual OECD countries. This assumption entails that in the 2030 target state, the share of RE fluctuating over time is no higher than 35%—which largely ensures that by 2030 no major investments will be required in storage facilities, DSM, additional electricity consumers, etc. so as to integrate volatile RE electricity into the electricity system.

7. Finally, the stated costs do not take into account the need for auxiliary financial resources to maintain the proportion of 50% of CO_2-free energy sources in the electricity mix well beyond 2030. The main reason for this simplification is this: Of course, any electricity generation system, i.e. any existing fleet of power plants, needs to be maintained, which demands regular new investment. Whether it is more or less expensive beyond 2030 to maintain today's electricity generation system or the one after the energy transition envisaged here is achieved is uncertain. In particular, it is not reliably predictable whether in a full cost calculation (i.e. comparing the cost for electricity from a newly constructed conventional power plant against those from a renewable energy power plant including the corresponding system technologies), fossil fuel power plants will be more or less expensive than renewable power plants after 2030.

Taking these assumptions as a basis, the following estimates can be derived for OECD countries.

22.5.2 Costs for a Typical OECD Country

Suppose a typical OECD country with the following electricity mix:

Nuclear energy	10%
Hydropower	5%
Other RE (PV, wind, etc.)	10%
Fossil fuels	75%

along with a gross electricity consumption of 300 TWh, an electricity efficiency of $ 4.8 per kWh (this corresponds to the average of all OECD countries) and thus a GDP of $ 1440 billion. Then for the energy transition according to our definition, (25% =) 75 TWh of RE electricity needs to be added until 2030, replacing a third of the fossil fuels and cutting CO2 emissions accordingly.

The total cost of this RE expansion (i.e. the necessary net subsidies) will be

$$ \$ \, 0.07 \text{ per kWh} \times 75 \text{ TWh} \times 20 = \$ \, 105 \text{ billion.} $$

This $ 105 billion must be paid in the period from 2017 to 2050 at an average of approx. $ 3 billion per year, equivalent to an average share of roughly 0.2% of the country's GDP over these years. If we raised the bar for the energy transition to a 66%-decarbonization of the electricity mix by 2030 (i.e., 125 TWh of RE to be added), the country would have to spend approx. 0.3% of its GDP over the next decades.

Table 22.1 Roughly estimated cost of an energy transition to 50%/66% share of CO2-free electricity, based on the electricity mix in 2015

Country	Nuclear energy (%)	Hydropower (%)	Other RE (%)	Electricity efficiency ($/kWh)	Costs (% of GDP) 50%/66%
USA	19	6	8	4.2	0.14/0.28
UK	19	3	18	7.7	0.05/0.12
Australia	0	7	8	4.4	0.28/0.41

https://data.oecd.org, own calculations

22.5.3   Costs for Individual OECD Countries

We can illustrate the same point by looking at a few examples of individual OECD countries with quite different electricity systems: the USA, UK and Australia. Using the calculation schema above, the Table 22.1 lists the roughly estimated costs of the envisaged energy transition up to a 50% (or 66%) share of CO2-free electricity for these countries, expressed as the percentage of the costs of the GDP.

We *do not* provide these figures here to really give a prediction of the costs of an energy transition in these countries. Rather, we want to highlight that it is quite probable that the costs of such a specific transformation will—if distributed over 3 decades—be in the range of only a few thousandths of the GDP in many OECD countries. This holds true even if the target for CO_2-free electricity in 2030 is increased to 66% or if the target was to replace half of the fossil fuels in electricity generation until 2030.

22.5.4   Conclusion

In other words, **an energy transition should in fact be affordable, at least for many countries in the OECD.** For example, if the energy transition were half-funded by private households (whose consumer spending averages approx. 60% of GDP in the OECD), in most countries these private households would also be burdened with no more than a few thousandths of their consumption spending.

> From today's perspective, energy transitions do not represent an excessive financial challenge for many OECD countries—at least up to a speed of development where CO_2-free energy sources should account for 50% or even 66% of the electricity mix by 2030.
>
> Indeed, energy transitions at this speed can typically be funded with a mere few thousandths of GDP. In particular, the expected burden on private households in these countries can be deemed not unreasonable.

Part IV

The German Energy Transition: Lessons to Be Learned

Answers to the Key Questions

The central purpose of this book is to give a systematic, impartial account and thus to create *transparency on the German energy transition* with respect to the electricity sector: transparency concerning the targets, motives and principles of the energy policy underlying the *Energiewende*, transparency about the objective current status of the *Energiewende*, transparency concerning major macroeconomic effects of the *Energiewende* and transparency about what could have been done better in the design and in the implementation of the *Energiewende*. Up to this point, our aim was to compile the essential facts and central arguments in the context of the *Energiewende*, to arrange them in a clear structure and, as far as possible, to derive further conclusions and judgements.

In this last part of the book we now want to use the base thus established—these facts, arguments and conclusions—to highlight the lessons and the insights likely to be the most interesting from an international point of view; i.e. to answer the five questions we posed in the introduction to this book (Sect. 1.1).

In deference to readers' differing time budgets and needs, we will do so in two different forms for each topic or question:

- *In one sentence* as a short answer to the question
- *In one section* for a more detailed discussion of the question

As mentioned before, here we basically draw on the data provided in the first three parts of the book. Occasionally, however, we will also introduce new facts and figures to round out the overall picture for a particular question.

Five Answers

23

23.1 Is the Energy Transition Too Expensive for the National Economy?

- In Germany, the *Energiewende* up until 2030 can be expected to require nominally € 600–700 billion in macroeconomic costs, which will accrue and have to be paid between 2000 and 2050. This equals an average of 0.4–0.5% of GDP annually over these 50 years. However, this money remains (for balance) within the national economy and so primarily causes internal redistributions.
- These costs are heavily affected by the fact that Germany has decided to phase out the CO_2-free nuclear energy until 2022 despite ambitious climate targets. Without this phase-out of nuclear energy, the costs (if the *Energiewende* targets were otherwise unchanged) would only be in the region of € 300–400 billion—and thus annually at an average of less than 0.25% of GDP.
- This result can essentially be translated to similar countries (such as the majority of OECD countries). An energy transition striving for 50% CO_2-free electricity supply in 2030 would probably cost just a few thousandths of GDP annually in many OECD countries.

We use the term "macroeconomic costs" in relation to the *Energiewende* to mean primarily the (net) subsidies that the national economy has to provide for the expansion of renewable energies (RE). The other negative and positive direct financial effects which can be ascribed directly to the *Energiewende* (grid expansion, lower electricity prices on the energy exchange, avoided construction of conventional power plants, subsidies for CHP plants etc.) are much lower and partly offset each other. In terms of the implementation of the *Energiewende* up to 2030—i.e. up

© Springer-Verlag GmbH Germany 2017
T. Unnerstall, *The German Energy Transition*, DOI 10.1007/978-3-662-54329-0_23

to the point at which at least half of the electricity is expected to come from renewable energies, marking about two-thirds of the way to the 2050 target for the *Energiewende*—these costs are relatively reliable to estimate. They can be expected to be in the region of € 600–700 billion, to be paid in the period 2000–2050, and mostly in the 25 years between 2010 and 2035.

This amount is nominally far higher than past subsidies for nuclear energy (estimated at approximately € 150 billion, excluding future storage costs) and for the German hard coal industry (estimated at approximately € 280 billon). Indeed, the costs of the *Energiewende* have already been pushing the German media to generate headlines such as "unbelievably expensive", "cost explosion in the *Energiewende*" and so on.

To properly assess these figures, however, a number of aspects should be considered:

- The trade balance of Germany's national economy remains mostly unaffected, i. e. these monies remain within the national economy. They are used in particular to pay revenues of German wind and PV companies, credit interest, returns on investments in the RE plants and finally operating costs for servicing and maintenance of the RE plants, insurance, rents for the requisite land, etc.
- Assuming moderate real economic growth of 1% per year in the future, the macroeconomic costs can be expected to account for on average 0.4–0.5% of GDP per year, over 50 years.
- The underlying investments in RE plants of approximately € 300 billion (2000–2030), or on average € 10 billion per year, account for only about 2% of the total German investment in fixed assets.
- It is true that energy costs to German consumers—private households, industry and others—have increased by more than € 100 billion, from € 130–140 billion per year in the 1990s to around € 250 billion per year in recent years. However, this is *not* primarily a consequence of the *Energiewende*, which accounts for roughly 20% of this increase. Rather the main reasons are higher global prices for oil, hard coal and natural gas (these prices alone amount to a plus of € 60 billion per year) alongside higher energy taxes.

For these reasons, the widespread view in Germany that the *Energiewende* is "too expensive" is not convincing. *Germany can afford the energy revolution.*

In this context and for comparison purposes, we should list the costs for some other social projects, relative to the size of the economy, i.e. relative to the GDP:

- (Direct) subsidies for German hard coal—emerging from the motive of reducing dependence on energy imports from abroad—between 1970 and 2010 totalled an estimated € 280 billion, equaling an annual average share of GDP of 0.3% over these 40 years.
- From 1991 to 2010, conservative estimates put the cost of German reunification at at least € 600 billion, equivalent to a share of on average 1.2% of the German GDP.

- The American Apollo programme—developed from the motive to be the first nation to land a man on the moon—between 1961 and 1972 costs roughly $ 24 billion, at the time equivalent to a share of on average 0.2% of the US GDP.
- All OECD countries have pledged to continuously spend 0.7% of their GDP on development aid in order to support less developped countries.

With respect to the costs of the *Energiewende*, there is one other important argument often invoked by its proponents: The economic balance of the *Energiewende* will be positive, so the argument goes, because by 2050 the value of fossil fuel imports avoided will be much higher than the macroeconomic cost of the *Energiewende*.

Unfortunately, here again we cannot share this view. The argument is based on the key assumption that the global prices for oil, hard coal and natural gas will continue to rise significantly and steadily until 2050, as they did in the past few decades (up to 2010/2011). In recent years, however, these prices *have dropped significantly*, and in our opinion their future development is completely open.

From an **international perspective**, the costs of the German energy transition are above all interesting in terms of the extent to which they can be translated to similar energy transitions, particularly in developped countries. The corresponding analysis clearly demonstrates that they *cannot* be translated.

The costs of the German energy transition are very significantly affected by the fact that Germany has taken and is still taking a quite particular path. Not only has Germany pioneered the expansion of renewable energies (especially PV) over the last decade, it is simultaneously proceeding with the expansion of RE *and* the phase-out of CO_2-free nuclear power. The particularity of this path is responsible for around 75% of the costs. In other words: If Germany had postponed the phase-out of its nuclear power plants (for about 15 years) and had started the massive expansion of renewable energies only in 2015, then the costs—targets and instruments being equal—would likely have been in the order of an estimated € 150–200 billion spread over 35 years, i.e. less than 0.15% of GDP.

This result can essentially be translated to other comparable countries.

If a country

- sets itself a target for 2030 similar to the one in the German energy transition (50% CO_2-free electricity by expanding mainly PV and wind power),
- continues to use nuclear energy (in case it is being used at present),
- and employs a funding mechanism that spreads the necessary subsidies for the RE plants over 20 years,

the costs of transforming an electricity system do not present an excessive financial burden on the national economy. It is true that these costs depend very heavily on the existing share of CO_2-free electricity in the electricity mix, on the natural conditions for renewable energies, on the electricity efficiency already achieved and on a variety of other country-specific parameters. However a rough but reasonable estimate shows that the costs are likely not to exceed around 0.2% of the country's GDP over a period of approx. three decades. Moreover, even for countries with a particularly low share of CO2-free electricity today—e.g. Australia or Mexico—these costs will probably not exceed 0.3–0.4% of GDP.

23.2   Can Private Households and Businesses in Germany Cope with the Financial Burdens of the Energy Transition?

> – Private households in Germany currently pay on average € 20–22 (including VAT) per month for the *Energiewende*. This is far below 1% of the total expenditure by private households, and far below consumer spending on items such as cigarettes or alcoholic beverages.
> – Germany's *non-energy-intensive* industries are currently burdened by the *Energiewende* with less than 0.5% of their sales volumes; by contrast, labour costs amount to roughly 20%.
> – The energy costs of the *energy-intensive industries* have not been increased by the *Energiewende*.

Currently, the overall burden on **private households** in Germany from the *Energiewende* is some € 10 billion per year (= GREA apportionment—electricity price advantage + grid costs + minor effects + VAT). It will continue to rise in the next few years but in all likelihood will not exceed € 13–15 billion per year. This must be compared against spending on, e.g. tobacco products (> € 20 billion per year), alcoholic beverages (> € 20 billion per year) or against the increase in real wages, e.g. in 2015 (> € 40 billion).

Considering these figures and a share of significantly less than 1% of total consumer spending, generally speaking we certainly cannot talk about a large or even undue burden being placed on private households by the *Energiewende*.

This being established, we must admit that it does not yet answer the question of a fair burden-sharing *between* the private households in Germany. Here, doubts are justified: the GREA surcharge does not respect the economic situation of a household. And even though we are not talking about large sums of money (€ 20–22 per month), this clearly is not an optimal solution for the issue of distributing the costs of the *Energiewende* within the national economy.

The additional costs for electricity that the *Energiewende* balance requires today (for balance) of the **German businesses** amount to approx. € 10 billion, broken down to:

– Industry, not energy intensive: approx. € 6 billion
– Trade and services sector: approx. € 4 billion
– Industry, energy intensive: 0

The trade and services sector generally is not subject to international competition. So if one primarily considers the non-energy-intensive industry that is subject to international competition (e.g. mechanical engineering, automotive etc.), the additional costs due to the *Energiewende* of around € 6 billion per year amount to a share of 0.4% of the sales volumes of these industries, which currently stand at roughly € 1,500 billion per year. While these costs are indeed noticeable, generally speaking it is not convincing to claim that German industry is suffering in terms of international competitiveness. For example, a single wage increase of 2% means a burden in the region of € 6 billion per year, i.e. on a similar scale.

This general statement does not exclude, of course:

– That an individual business whose the energy costs are well above the average of 1.5% of sales volumes for non-energy-intensive industries (but that are nevertheless not classified as "energy intensive" as defined by the GREA; energy costs for energy-intensive businesses are on average 5–6% of sales volumes), might actually suffer from the additional costs resulting from the *Energiewende*. The statement is only that this can be considered as an individual case.
– That (at least for some particularly energy-intensive sectors of German industry) a number of German production sites do have a significant competitive disadvantage (e.g. compared to the USA) due to the energy costs in general. The statement is only that this disadvantage has essentially nothing to do with the *Energiewende*.

In this context let us note that it is, however, indispensable for Germany's energy-intensive industries to remain (largely) exempt from the additional costs resulting from the *Energiewende*.

23.3   What Are the Actual Effects of the Energy Transition on CO_2 Emissions?

From today's perspective, the central motive of the *Energiewende*—to drastically reduce CO_2 emissions and thus fight climate change—will be achieved, albeit quite slowly, in terms of the *German* CO_2 emissions.

However, the extent to which the *Energiewende* has a noticeable influence on *global* CO_2 emissions and thus on climate change will depend on the extent to which relevant technologies, political concepts and lessons learned can and will be exported to other countries. Today, this is an open question.

Internationally, the clear majority of scientists hold that anthropogenic CO_2 emissions are the main cause of global climate change as we currently perceive it, and they advocate for the 2 °C goal, i.e. the demand that humanity must limit the

Table 23.1 German CO_2 emissions from electricity generation (excluding electricity exports) (in mio t)

Year	CO_2 emissions
1990	360
2000	330
2010	305
2015	270
2030	200 (predicted)
2050	40–60 (planned)

increase in global average temperature to a maximum of 2 °C above preindustrial levels in order to be able to reasonably manage the consequences of climate change.

In accordance with this conviction—as was stated at the G7 at Schloss Elmau in June 2015 and at the Paris Climate Conference in December 2015—the industrialized nations in particular are under the obligation to "decarbonize" their economies and societies, i.e. to drastically cut their CO_2 emissions in the next decades.

The German energy transition in fact pursues this as one of its central motives. It is intended to reduce energy-related CO_2 emissions in Germany from their original levels of 1,000 mio t in 1990 and 800 mio t in 2010 to 150–200 mio t in 2050. In the electricity sector, the goal is to reduce CO_2 emissions caused by electricity generation from the original levels of 360 mio t in 1990 and 305 mio t in 2010 to around 50 mio t in 2050.

This CO_2 reduction in the electricity sector due to the *Energiewende* will progress quite slowly, but from today's perspective—further consistent implementation provided—it will be achieved (Table 23.1; compare Table 12.4). The main reason for the slowness in this development lies in the simultaneous phase-out of CO_2-free nuclear power.

By contrast, the two other major energy sectors—heating and transportation—have realized virtually no reduction in CO_2 emissions since 2010. Given this and the significant resources spent on the *Energiewende*—still significant though affordable for the national economy—we are led to ask whether it might not have made more sense to drive the *Energiewende* in the transportation and heating sectors instead of in the electricity sector?

Exploring this question (see Sect. 16.4) proves that the answer has to be a resounding no. Although there are a number of options that are significantly cheaper than the current *Energiewende* (in the electricity sector) for reducing CO_2 emissions, namely, in the heating and transportation sectors, this potential is nowhere near sufficient to achieve the target state in 2050. In other words, the drastic reduction in CO_2 emissions (also) in electricity generation—i.e. the *Energiewende* as discussed in this book—is simply indispensable if CO_2 emissions are to be reduced in Germany to the extent as intended by the German government and as envisaged by the Paris Climate Agreement.

The truly relevant question relating to CO_2 emissions, however, is a different one: Does the *Energiewende* contribute to *global* climate protection at all?

There are certainly doubts to be taken seriously here. We found (Sect. 18.4) that the *Energiewende* up to 2014—which alone is responsible for approx. € 400 billion of the total € 600–700 billion in macroeconomic costs—has reduced CO_2 emissions in the German electricity sector only by about 60 mio tons per year compared to the year 2000. Over a period of 20 years, this corresponds to about two weeks' worth of global CO_2 emissions.

More drastically speaking: Germany is spending € 400 billion to delay climate change by just two weeks.

Obviously, this is—putting it mildly—not a particularly good ratio. It certainly is not if we assume that there are measures to reduce CO_2 *in other countries* that could be realized by Germany which offer significantly higher CO_2 cost-efficiency compared to the *Energiewende* (which is geared towards reducing CO_2 in Germany).

Thus, German policymakers must ask themselves: Would it not have made sense to use some of the significant resources and political efforts expended on the *Energiewende* to reduce CO_2 emissions in other countries and so achieve a faster and better impact?

This question cannot be dismissed, and it is probably the most important legitimate criticism of the *Energiewende* concept as well as being one of the most important challenges to future energy policy decisions in Germany.

Let us reword this once again: Currently, Germany is almost solely relying on its *Energiewende* being a technical, conceptual and maybe ethical *role model* as regards the global effects of its climate protection policy. No matter how we rate this role model function, i.e. the ability to export technologies, political concepts and lessons learned relating to the *Energiewende*, for the present and for the future: there has been no *systematic evaluation of alternative options* for Germany to affect CO_2 emissions in other countries more directly.

However, this question, this criticism and this call for systematic evaluation are directed not only to politicians, but also to science in its advisory capacity, to professional associations and, ultimately, to society as a whole. Moreover, they are directed not only to Germany, but to the other industrialized countries in the world as well.

In this context we must also concede that without a certain role model function—i.e. without a consistent, verifiable path for reducing CO_2 emissions in Germany—such efforts in other countries would be politically doomed to failure.

Consequently, this is not an either-or issue but rather concerns the question of prioritization and timing of international actions in relation to the current, almost exclusively domestically focused *Energiewende*. After all, the climate fund agreed at the global climate conferences in Copenhagen (2009) and Paris (2015)—initially € 100 billion per year from 2020—is an essential step in this direction by the industrialized countries.

23.4   What Lessons Can Be Learned from the Design of the Energy Transition?

If we take the basic concept of the German energy transition as well as the target state in 2050 it sets itself as an indisputable basis, there are three main lessons to be learned from the (weaknesses of the) specific design of the *Energiewende*. When designing an energy transition with ambitious long-term targets, it is important to:
- Place high emphasis on the principle of cost-efficiency, in particular with respect to the milestones on the way to a decarbonized electricity system.
- Not only set milestones for the expansion of renewable energies but also corresponding milestones for the necessary expansion of the grid.
- More generally to soberly try to foresee systemic effects along the way and to deal with them from the outset as far as possible.

The basic concept of the German energy transition in the electricity sector—to replace nuclear energy and in the long-term fossil fuels with renewable energies in electricity generation and to substantially increase electricity efficiency (or electricity productivity)—is a practically inevitable consequence of political motives which are deeply rooted in German politics and society. It can therefore, for Germany, be deemed indisputable. Further taking into account the current status of global climate policy and in particular the Paris Climate Agreement of 2015, it is equally indisputable that an industrialized nation like Germany must cut its CO_2 emissions and in particular those from electricity generation to a level close to zero by 2050.

In the design of the *Energiewende*, i.e. with respect to its key targets, the *target state of the German electricity system in 2050* is thus largely fixed and conceptionally sound:

- No nuclear energy
- At least 80% share of renewable energies
- Energy productivity doubled.

By contrast, the *milestones on the way to 2050* set in the *Energiewende* design of 2010/2011 appear to contain a considerable degree of political arbitrariness. These milestones set out an unnecessarily high speed for the structural change in the electricity system which—while technically feasible and also objectively affordable for Germany's national economy and its stakeholders—renders it unnecessarily expensive. In other words: They do not respect the principle of cost-efficiency, which is important if not crucial for the overall political and social consensus to last over the next decades—and without this consensus, in turn, such a long-term project will be doomed to failure.

Any energy transition in an industrialized country is likely to require a considerable number of new powerlines. At least from today's perspective, the principle of cost-efficiency entails that the main criterion for the choice of locations for larger RE plants should be "Where are the most favourable natural conditions for high electricity production?" rather than (as for conventional power plants) "Where are the centres of electricity consumption and where are existing grids we could use?". In general, the regions of a country with most favourable natural conditions for RE plants will not be the regions with high electricity consumption: big cities, industrial centres, etc. (this is particularly obvious with respect to wind offshore technology). Of course, the more specific circumstances in this respect will differ widely from country to country, but it is rather safe to say that new transmission grids will indeed be needed in the course of the expansion of RE, in order to connect the centres of RE electricity production with the centres of electricity consumption.

The German experience clearly shows that neglecting this basic relation or underestimating the time to build such new power lines leads to a whole host of complex and tedious problems and, in particular, to additional and systemically unnecessary costs.The best way to avoid such problems is to coordinate RE expansion and the foreseeable necessary grid expansion from the beginning, i.e. already in the design of the energy transition. In other words, the German experience suggests that this design should comprise not only milestones for RE expansion but also corresponding milestones for the ensuing grid expansion. And in the implementation of the energy transition, both sets of milestones should be monitored and politically enforced with the same emphasis.

Grid expansion, however, is actually only one of several systemic consequences—albeit a very important one—when an electricity system undergoes the fundamental structural transition from conventional power plants to RE power plants. On the one hand, it is impossible and even dangerous to try to predict *all* such effects for the *long-range future* and draw up a kind of "master plan" for the energy transition for the next decades—it is important to stay open to new technologies, to faster-than-expected cost degressions, to unforeseeable possibilities for cooperations with neighbouring countries and to many other possible options that might facilitate and render cheaper the energy transition. On the other hand, *some* of such systemic consequences pertaining to a *mid-range future* might be reliably foreseeable when launching (and in the course of) an energy transition, for example, developments in the conventional power plant fleet and ensuing consequences for the energy companies operating these plants, higher overall cost of the new electricity system and ensuing burdens on society and others.

Energy policy can proactively deal with such developments, mitigate possible negative effects on stakeholders or at least prepare the energy industry and the society in its public communications. Again, the German example shows that the failure to do so risks to lead to long, heated and bitter discussions as well as to hectic and shortsighted political actions. Or, to put it positively: proactive political management of foreseeable consequences for the electricity system in the course of gradually replacing conventional power plants with RE plants is a very helpful factor for a relatively smooth path of an energy transition.

23.5   What Lessons Can Be Learned from the Implementation of the Energy Transition?

> While overall the implementation of the *Energiewende* can be rated as successful so far, this success has come at an unnecessarily high price, and the German society was not well prepared to encounter these costs.
>
> This stems from weaknesses in its design (see 23.4), but it is also due to far-reaching mistakes in its implementation. The most important lesson from the German experience in this respect is that it is preferable to use a tender model in the expansion of renewable energies rather than state-fixed feed-in tariffs.

The implementation of the *Energiewende* can be rated as successful so far in most respects. The milestones on the way to the defined targets—successive shutdown of nuclear power plants, steady expansion of renewable energies and significant increase in electricity efficiency—have been achieved, even exceeded; the main motives—reducing CO_2 emissions, phasing out nuclear energy and decreasing dependence on (foreign) fossil fuels—have consequently been fulfilled to the planned degree; and two of the important framework conditions of German energy policy—security of supply and market economy in conventional electricity generation—have been complied with.

The core problem of the *Energiewende* implementation so far is that the third framework condition—to safeguard the affordability of energy by adhering to the principle of cost-efficiency—has not been complied with.

It is true—as we have pointed out (23.1, 23.2)—that, from an objective point of view, electricity is still well affordable for the German national economy and its main stakeholders on a whole as well as for most individual stakeholders. Nevertheless, the subjective perception is often different, most of the social and political controversy around the *Energiewende* pertains to cost issues, and the rising cost risks to threaten the indispensable social consensus on the *Energiewende*. Moreover, the substantial cost figures do tarnish the international reputation of the *Energiewende*—which is important when it comes to its possible positive impacts outside of Germany. For these reasons, cost-efficiency in the *Energiewende* (in any energy transition) is crucial.

Since the main cost factor in an energy transition is the required subsidies for the RE power plants, it is clear that strict adherence to the principle of cost-efficiency in energy policy is nowhere more important than when regulating these subsidies—i. e. for Germany, in the regulations provided by the GREA (German Renewable Energy Act).

For the past 15 years, the basic concept of the GREA in this respect was to define fixed feed-in tariffs (in ct/kWh) for the various RE types—PV, wind and biomass— i.e. to guarantee for the investor a fixed reimbursement for every kilowatt-hour of electricity produced within the first 20 years of operation. While, roughly speaking, in the first 10 years (2000–2010), this concept proved to be adequate and successful, in the past five years, it led to serious breaches of the principle of cost-efficiency: it led to an uncontrolled expansion of RE—too fast, too high a share of PV—and it led to too high ROIs for investors.

The key problem here is that in a dynamic, globally driven market of RE technology, feed-in tariffs fixed by law cannot be adjusted quickly enough to changing market conditions, in particular to cost degressions of RE plants within short periods of time: changing these figures in the GREA through a parliamentary process in a democracy is simply too slow.Moreover, in any case it is more in line with the principles of market economy to let not the state but to let the market decide on necessary subsidies and (minimum) required ROI for investors in RE power plants. This consideration can be implemented by the concept of tenders for certain volumes of RE capacity, organized by the state, where basically any investor can participate in free competition. The result is that the investors requiring the least subsidies for their projects are selected to realize their RE plants. Of course, there are a number of questions to be resolved concerning the details, but in principle this concept is clearly superior to the previous one—in terms of cost-efficiency and also in terms of controlling the speed of expansion of renewable energies.

Against this background, it must be rated a serious mistake that in Germany it took so long (until 2016) to switch over to this tender concept—a mistake that is responsible for unnecessary costs in the range of at least € 100 billion. (Just recently the power of the tender concept has been illustrated in a striking way: tenders for PV systems and for offshore wind farms in the first half of 2017 resulted in required subsidies much lower than anybody expected.)

From conceptual considerations as well as from the German experience, then, it is rather safe to conclude that the best way to keep the cost of an energy transition—from conventional power plants to renewable power plants—to the unavoidable minimum is to employ a tender concept for the construction of the RE power plants.

Finally, in view of this conclusion, in view of the answer to question 23.1 and in view of the increasing dynamics and global competition within the RE industry, we believe **there are rising chances that energy transitions all over the world can be politically designed, controlled and monitored in such a way that they are indeed affordable.**

Appendix

1. The figures and data used in this book are—with only a few exceptions—taken from or have been derived from the information given on the websites of the following organizations and institutions:
   - AG Energiebilanzen (AGEB)—http://www.ag-energiebilanzen.de
   - Bundesministerium für Wirtschaft und Energie (Ministry of Economic Affairs and Energy, BMWi)—http://www.bmwi.de
   - Bundesministerium für Umwelt (Ministry for the Environment, BMUB)—http://www.bmub.bund.de
   - Umweltbundesamt (Federal Environment Agency, UBA)—http://www.umweltbundesamt.de
   - Statistisches Bundesamt (Federal Statistical Office)—http://www.destatis.de
   - Bundesnetzagentur (Federal Grid Agency, BNA)—http://www.bundesnetzagentur.de
   - Bundesverband der Energie und Wasserwirtschaft (German Association of the Energy and Water Industries, BdEW)—http://www.bdew.de
   - Übertragungsnetzbetreiber (TSOs)—http://www.netztransparenz.de
   - European Energy Exchange (EEX, German Power Exchange)—http://www.eex.com
   - International Energy Agency (IEA)—http://www.iea.org
   - Organisation for Economic Cooperation and Development (OECD)—http://www.oecd.org
   - Fraunhofer ISE, Energy Charts—http://www.energy-charts.de
   - Weltenergierat (World Energy Council)—http://www.weltenergierat.de.
2. The following documents, which can mostly be found on the cited websites, have been of particular importance in the process of writing this book:
   - Leadstudy 2011 (Langfristszenarien und Strategien für den Ausbau der erneuerbaren Energien in Deutschland, 29.03.2012)—http://www.dlr.de/dlr/Portaldata/1/Resources/bilder/portal/portal_2012_1/leitstudie2011_bf.pdf
   - AGEB (2016): Evaluation Tables on the Energy Balance 1990 to 2015
   - AGEB (2016): Stromerzeugung nach Energieträgern 1990–2016
   - AGEB (2016): Energy Consumption in Germany in 2015

© Springer-Verlag GmbH Germany 2017
T. Unnerstall, *The German Energy Transition*, DOI 10.1007/978-3-662-54329-0

- BMWi (2013): Energie in Deutschland—Trends und Hintergründe zur Energieversorgung, 2013
- BMWi (2014): Employment from renewable energy in Germany: expansion and operation—now and in the future, 2014
- UBA (2016): Entwicklung der spezifischen CO_2-Emissionen des deutschen Strommix in den Jahren 1990–2015
- UBA (2016): Submission under the United Nations Framework Convention on Climate Change and the Kyoto Protocol 2016
- UBA (2016): Treibhausgasemissionen in Deutschland seit 1990 nach Gasen
- BNA (2016): Kraftwerksliste November 2016 (List of power plants)
- BdEW (2016): Erneuerbare Energien und das EEG: Zahlen, Daten, Grafiken (2016)
- BdEW (2015): Bio-Erdgas: Fragen, Antworten und Argumente
- ÜNB (2016): Konzept der ÜNB zur Prognose und Berechnung der EEG-Umlage 2016, 2017
- BDI (2007): Kosten und Potenziale der Vermeidung von Treibhausgasemissionen in Deutschland—2007/2009
- DIW-Wochenbericht Nr.41/2010
- Weltenergierat-Deutschland e.V: Energie für Deutschland 2015, 2016
- BAFA, Aufkommen und Export von Erdgas sowie die Entwicklung der Grenzübergangspreise ab 1991

Zeitfracht Medien GmbH
Ferdinand-Jühlke-Straße 7
99095 Erfurt, Deutschland
produktsicherheit@kolibri360.de